Caroline Indorf

Aminozucker - Biomarker für Mikroorganismen; Methoden und Techniken

Caroline Indorf

Aminozucker - Biomarker für Mikroorganismen; Methoden und Techniken

Chromatographie der Aminozucker mittels HPLC sowie HPLC-IRMS

Südwestdeutscher Verlag für Hochschulschriften

Impressum/Imprint (nur für Deutschland/only for Germany)
Bibliografische Information der Deutschen Nationalbibliothek: Die Deutsche Nationalbibliothek verzeichnet diese Publikation in der Deutschen Nationalbibliografie; detaillierte bibliografische Daten sind im Internet über http://dnb.d-nb.de abrufbar.

Alle in diesem Buch genannten Marken und Produktnamen unterliegen warenzeichen-, marken- oder patentrechtlichem Schutz bzw. sind Warenzeichen oder eingetragene Warenzeichen der jeweiligen Inhaber. Die Wiedergabe von Marken, Produktnamen, Gebrauchsnamen, Handelsnamen, Warenbezeichnungen u.s.w. in diesem Werk berechtigt auch ohne besondere Kennzeichnung nicht zu der Annahme, dass solche Namen im Sinne der Warenzeichen- und Markenschutzgesetzgebung als frei zu betrachten wären und daher von jedermann benutzt werden dürften.

Coverbild: www.ingimage.com

Verlag: Südwestdeutscher Verlag für Hochschulschriften GmbH & Co. KG
Heinrich-Böcking-Str. 6-8, 66121 Saarbrücken, Deutschland
Telefon +49 681 37 20 271-1, Telefax +49 681 37 20 271-0
Email: info@svh-verlag.de

Zugl.: Witzenhausen, Universität Kassel, Diss., 2011

Herstellung in Deutschland (siehe letzte Seite)
ISBN: 978-3-8381-3332-4

Imprint (only for USA, GB)
Bibliographic information published by the Deutsche Nationalbibliothek: The Deutsche Nationalbibliothek lists this publication in the Deutsche Nationalbibliografie; detailed bibliographic data are available in the Internet at http://dnb.d-nb.de.

Any brand names and product names mentioned in this book are subject to trademark, brand or patent protection and are trademarks or registered trademarks of their respective holders. The use of brand names, product names, common names, trade names, product descriptions etc. even without a particular marking in this works is in no way to be construed to mean that such names may be regarded as unrestricted in respect of trademark and brand protection legislation and could thus be used by anyone.

Cover image: www.ingimage.com

Publisher: Südwestdeutscher Verlag für Hochschulschriften GmbH & Co. KG
Heinrich-Böcking-Str. 6-8, 66121 Saarbrücken, Germany
Phone +49 681 37 20 271-1, Fax +49 681 37 20 271-0
Email: info@svh-verlag.de

Printed in the U.S.A.
Printed in the U.K. by (see last page)
ISBN: 978-3-8381-3332-4

Copyright © 2012 by the author and Südwestdeutscher Verlag für Hochschulschriften GmbH & Co. KG and licensors
All rights reserved. Saarbrücken 2012

Die Wissenschaft, richtig verstanden, heilt den Menschen von seinem Stolz:
denn sie zeigt ihm seine Grenzen.

(Albert Schweitzer)

Vorwort

Die vorliegende Dissertation wurde im Rahmen des DFG- Projektes JO 362/7 an der Universität Kassel im Fachbereich Ökologische Agrarwissenschaften im Fachgebiet Bodenbiologie und Pflanzenernährung erstellt, um die Anforderungen des akademischen Grades Doktor der Naturwissenschaften zu erfüllen. Die Arbeit beinhaltet 3 Artikel, von denen einer bereits bei einer international, begutachteten Fachzeitschrift veröffentlicht wurde. Der zweite und der dritte Artikel wurden im Dezember 2011 eingereicht. Diese Artikel sind in Kapitel 3, 4 und 5 eingebettet. Kapitel 1 ist eine generelle Einleitung zum Thema, während Kapitel 2 die Ziele dieser Arbeit beschreibt. In Kapitel 6 und 7 werden die Ergebnisse aus Kapitel 3, 4 und 5 auf Deutsch sowie auf Englisch zusammenfassend dargestellt. Kapitel 7 gibt einen Ausblick auf weitere Untersuchungen, während in Kapitel 8 die Quellen der Kapitel 1, 2 und 7 aufgeführt sind.

Die folgenden Artikel sind in diese Arbeit eingearbeitet worden:

Kapitel 3

Indorf, C., Dyckmans, J., Khan, K.S., Joergensen, R.G. (2011). Optimisation of amino sugar quantification by HPLC in soil and plant hydrolysates. Biol Fertil Soils (DOI 10.1007/s00374-011-0545-5).

Kapitel 4

Indorf, C., Bodé, S., Boeckx, P., Dyckmans, J., Meyer, A., Fischer, K., Joergensen, R.G.

Comparison of HPLC methods for determination of amino sugars in soil hydrolysates. Soil Biology and Biochemistry, submitted.

Kapitel 5

Indorf, C., Stamm, F., Dyckmans, J., Joergensen, R.G.

Determination of saprotrophic fungi turnover in different substrates by glucosamine specific $\delta^{13}C$ liquid chromatography/isotope ratio mass spectrometry. Fungal Ecology, submitted.

Inhaltsverzeichnis

Abbildungsverzeichnis

Tabellenverzeichnis

Abkürzungsverzeichnis

1. Einleitung 1
   1.1   Vorkommen und Bedeutung von Aminozuckern 1
   1.3   Aminozuckeranalytik – Chromatographie der Aminozucker 5
   1.4   Aminozuckeranalytik – Isotopenmassenspektrometrie 7
2. Ziele der Arbeit 10
3. Optimisation of amino sugar quantification by HPLC in soil and plant hydrolysates 11
   3.1. Introduction 12
   3.2. Materials and methods 13
       3.2.1. Soil and plant samples 13
       3.2.2. Solutions 15
       3.2.3 Chromatographic conditions 15
   3.3. Results 16
       3.3.1 Method optimisation 16
   3.4 Discussion 24
       3.4.1 Optimisation of the mobile phase 24
       3.4.2 Excitation wavelength optimisation 25
       3.4.3 Ortho-phthaldialdehyde (OPA) reaction time 26
       3.4.4 Validation parameters 26
       3.4.6 Concentrations of amino sugars in soil and plant material samples 27
   3.5   Conclusions 28
   3.6 References 28
4. Comparison of HPLC methods for determination of amino sugars in soil hydrolysates 32
   4.1   Introduction 33
   4.2   Materials and methods 34
       4.2.1   Soils samples 34
       4.2.2   Chemicals 34
       4.2.3   Amino sugar extraction 36

4.2.4	Purification and concentration of amino sugars	36
4.2.5	Reversed phase-HPLC and pre-column derivatisation	38
4.2.6	HPAEC and post column derivatisation	38
4.2.7	HPAEC and pulsed amperometric detection (PAD)	42
4.2.8	HPCEC and post-column derivatisation	42
4.2.9	HPEXC and post column derivatisation	43
4.3	Results	44
4.3.1	Sample purification methods	44
4.3.2	Chromatographic methods	45
4.3.2.1	*HPAEC analysis with fluorescence detection*	45
4.3.2.2	*HPAEC analysis and pulsed amperometric detection*	50
4.3.2.3	*HPCEC and HPEXC analysis and fluorescence detection*	51
4.4	Discussion	55
4.4.1	Sample purification methods	55
4.4.2	HPAEC-Fl	55
4.4.3	HPAEC-PAD	56
4.4.4	HPCEC and HPEXC	57
4.4.5	Comparison of RP-Fl, HPAEC-Fl and HPAEC-PAD	58
4.5	Conclusions	59
4.6	References	60

5. Determination of saprotrophic fungi turnover in different substrates by glucosamine-specific $\delta^{13}C$ liquid chromatography/isotope ratio mass spectrometry 63

5.1	Introduction	64
5.2	Materials and methods	65
5.2.1	Material	65
5.2.2	Solutions	66
5.2.3	Substrate preparation and spawning	66
5.2.4	Amino sugar extraction	67
5.2.5	Amino sugar determination	67
5.2.6	Amino sugar specific $\delta^{13}C$ analysis by HPAEC-IRMS	68
5.2.7	Total C and total $\delta^{13}C$	68
5.2.8	Calculations and statistical analysis	68
5.3	Results	69
5.3.1	HPAEC-IRMS method optimisation	69

    5.3.2   The fungal growth experiment 70
  5.4   Discussion 75
    5.4.1   HPAEC-IRMS optimisation 75
    5.4.2   The fungal growth experiment 76
  5.5   Conclusions 77
  5.6   References 78
6. Zusammenfassung 82
7. Summary 86
8. Schlussfolgerung und Ausblick 90
9. Literatur 92

Abbildungsverzeichnis

Abb. 1	Strukturformeln der Aminozucker-Monomere	3
Abb. 2	Baueinheit von Chitin	4
Abb. 3	Aufbau des Peptidoglycans (Murein) gramnegativer Bakterien	4
Abb. 4	a) Prinzip des Trennmechanismus Kationenaustauschchromatographie, b) Anionenaustauschchromatographie	6
Abb. 5	Reaktionsmechanismus der OPA- Derivatisierung	6
Abb. 6	a) Prinzip des Trennmechanismus Umkehrphasen(RP)-Chromatographie, b) Säulenmaterial Umkehrphasen(RP)-Chromatographie	7
Abb. 7	^{13}C-Isotopengehalt der wichtigsten Kohlenstoffreservoire der Erde (Carle, 1991)	9
Abb. 8	(Fig. 1) Chromatograms of (a) a standard mixture consisting of 13µmol l^{-1} muramic acid, 130 µmol l^{-1} mannosamine, 130 µmol l^{-1} galactosamine, 130 µmol l-1 glucosamine and of (b) a soil hydrolysate	18
Abb. 9	(Fig. 2) Chromatograms of (a) an amino sugar standard mixture consisting of 17 µmol l^{-1} muramic acid, 170 µmol l^{-1} mannosamine, 170 µmol l^{-1} galactosamine, 170 µmol l^{-1} glucosamine and of (b) an amino acid standard 1, (c) an amino acid standard 2 and (d) an amino sugar standard spiked with an amino acid standard 2, components of amino acid standards (see materials and methods).	19
Abb. 10	(Fig. 3) Optimisation of HPLC parameters: (a) sensitivity of the amino sugar determination as a function of OPA reaction time; (b) optimisation of the excitation wavelength; (c) changes in area as a function of eluent pH value; (d) Change in resolution between muramic acid (MurN) and mannosamine (ManN) peaks and in retention time of MurN, ManN, galactosamine (GalN) and glucosamine (GlcN) as a function of tetrahydrofuran (THF) concentration in eluent (error bars represent ± standard deviation).	21
Abb. 11	(Fig. 4) Calibration curves of reference amino sugars (error bars represent ± standard deviation)	22
Abb. 12	(Fig. 1) Comparison of four different sample purification methods	45
Abb. 13	(Fig. 2) Chromatograms of a standard mixture consisting of 130 µmol l^{-1} GalN, 130 µmol l^{-1} ManN, 130 µmol l^{-1} GlcN and 13 µmol l^{-1} MurN (a) obtained from the RP-Fl method and (b) obtained from the HPAEC-PAD method; chromatograms of (c) a standard mixture consisting of 130 µmol l^{-1} GalN, 130 µmol l-1 ManN, 130 µmol l-1 GlcN measured by HPAEC-Fl for basic amino	

	sugars and of (d) a standard solution containing 13 µmol l^{-1} muramic acid obtained from the acidic HPAEC-Fl method.	46
Abb. 14	(Fig. 3) Chromatograms of a soil hydrolysate (forest 1) (a) measured by RP-Fl, (b) obtained from the HPAEC-PAD method (c) obtained from the HPAEC-Fl method for basic amino sugars and (d) obtained from the acidic HPAEC-Fl method.	47
Abb. 15	(Fig 4) Calibration curves of reference amino sugars obtained from HPAEC-Fl	50
Abb. 16	(Fig. 5) Calibration curves of reference amino sugars obtained from HPAEC-PAD	51
Abb. 17	(Fig. 1) Chromatograms of a standard mixture consisting of 200 µmol l^{-1} GalN and 200 µmol l^{-1} GlcN (a) obtained from the HPAEC-IRMS method based on Bodé *et al.* (2009) and (b) obtained from the optimised HPAEC-IRMS method (c) a wheat straw hydrolysate measured by HPAEC-IRMS method described by Bodé *et al.* (2009) and (d) a wheat straw hydrolysate obtained from the optimised HPAEC-IRMS method.	71
Abb. 18	(Fig. 2) Formation of new fungal C in wheat-wood and maize-wood samples inoculated with fungi during 4 week incubation at 24°C	74
Abb. 19	(Fig. 3) Formation of new maize derived GlcN in maize-wood samples inoculated with fungi during 4 week incubation at 24°C	74

Tabellenverzeichnis

Tabelle 1	(Table 1) Physical, chemical and microbiological properties of the soils used in this investigation	13
Tabelle 2	(Table 2) Chemical properties of the plant litter materials used in this investigation	14
Tabelle 3	(Table3) Validation parameters of the HPLC method	22
Tabelle 4	(Table 4) Effect of sample solvent on the contents of muramic acid, mannosamine, galactosamine and glucosamine in soil samples	23
Tabelle 5	(Table 5) Effect of sample solvent on the contents of muramic acid, mannosamine, galactosamine and glucosamine in plant litter samples	23
Tabelle 6	(Table 1) Physical, chemical and microbiological properties of the soils used in this investigation	35
Tabelle 7	(Table 2) Chromatographic conditions of the different methods	40
Tabelle 8	(Table 3) Gradient profile for amino sugar separation by HPAEC-PAD	42
Tabelle 9	(Table4a) Validation parameters of the RP-Fl, HPAEC-Fl and HPAEC-PAD methods	48
Tabelle 10	(Table 4b+c) Validation parameters of the RP-Fl, HPAEC-Fl and HPAEC-PAD methods	49
Tabelle 11	(Table 5a+b) Comparison of amino sugar results in soil samples obtained by RP-HPLC, HPAEC-Fl and HPAEC-PAD	53
Tabelle 12	(Table 1) HPAEC-IRMS method optimisation	65
Tabelle 13	(Table 2) Chemical properties of the material used in the fungus experiment	66
Tabelle 14	(Table 3) Total C in wheat-wood and maize-wood mixtures inoculated with *Lentinula edodes*, *Pleurutus ostreatus*, and *Pleurotus citrinopileatus*, respectively, during 4 weeks incubation	72
Tabelle 15	(Table 4) Glucosamine (GlcN) amounts and GlcN-specific $\delta^{13}C$ values in wheat (W) and maize (M) samples inoculated with *Lentinula edodes* (LE), *Pleurutus ostreatus* (PO) and *Pleurotus citrinopileatus singer* (PC), respectively, during 4 week incubation	73

Abkürzungsverzeichnis

ANOVA	Varianzanalyse
$BaCl_2$	Bariumchlorid
$BaCO_3$	Bariumcarbonat
C	Kohlenstoff
^{13}C	Kohlenstoffisotop mit der Masse 13
$CaCl_2$	Calciumchlorid
CO_2	Kohlenstoffdioxid
CV	Variationskoeffizient
DFG	Deutsche Forschungsgemeinschaft
EA	Elementaranalysator
Fl	Fluoreszenz
g	Gramm
GalN	Galaktosamin
GC	Gaschromatographie
GC-C-IRMS	Gaschromatographie-Verbrennungs-Isotopenmassenspektrometrie
GlcN	Glucosamin
Gt	Gigatonne
H_2O	Wasser
H_3BO_3	Borsäure
H_3PO_4	Phosphorsäure
HCl	Salzsäure
HNO_3	Salpetersäure
HPAEC	Hochleistungsflüssigkeitanionenaustauschchromatographie
HPCEC	Hochleistungsflüssigkeitkationenaustauschchromatographie
HPEXC	Hochleistungsflüssigkeitanionenausschlusschromatographie
HPLC	Hochleistungsflüssigkeitchromatographie
KH_2PO_4	Kaliumdihydrogenphosphat
KOH	Kaliumhydroxid
IRMS	Isotopenverhältnismassenspektrometrie
LC	Flüssigkeitchromatographie
LOD	Nachweisgrenze

LOQ	Bestimmungsgrenze
M	Molar
ManN	Mannosamin
mg	Milligramm
µg	Mikrogramm
min	Minute
ml	Milliliter
µl	Mikroliter
mmol	Millimolar
µmol	Mikromolar
MurN	Muraminsäure
mV	Millivolt
N	Stickstoff
Na	Natrium
Na_2HPO_4	Natriumhydrogenphosphat
NaOH	Natronlauge
$NaNO_3$	Natriumnitrat
$Na_2S_2O_8$	Natriumpersulfat
$(NH_4)_2SO_4$	Ammoniumsulfat
ODS	Octadecylsilan
OH^-	Hydroxidion
OPA	ortho-Phthaldialdehyd
PAD	gepulster amperometrischer Detektor
PDB	PeeDeeBelmnite
pH	negativer dekadischer Logarithmus der Wasserstoffionenaktivität (*potentia Hydrogenii*)
R-	organischer Rest
r	Korrelationskoeffizient
RP	Umkehrphase
RSA	relative spezifische Allokation
s	Sekunde
SAX	starker Anionenaustauscher
SCX	starker Kationenaustauscher
SD	Standardabweichung

SOC	organischer Kohlenstoff des Bodens
SO_3^-	Sulfonsäureion
WHC	Wasserhaltekapazität

1. Einleitung

1.1 Vorkommen und Bedeutung von Aminozuckern

Aminozucker sind als Bestandteile von Zellwänden für viele Organismen von Bedeutung. So sind sie in der bakteriellen Zellwand zu finden, und in Pilzen sowie in Invertebraten liefern sie den Baustein für das strukturbildende Chitin. Aminozucker sind bis zu einem gewissen Maße aber nicht nur in der Zellwand sondern auch in anderen Zellkompartimenten nachgewiesen worden (Parsons, 1981). Bei höheren Lebewesen kommen sie insbesondere als Baustein in Mucopolysacchariden vor. Mucopolysaccharide sind Bestandteile vieler Makromoleküle. So sind sie als Faserstoffe in Zellmembranen aber auch im Blut oder Frauenmilch zu finden (Ternes et al., 2007).

Besondere Bedeutung gebührt den Aminozuckern jedoch in den Bodenwissenschaften, wo sie als Indikatoren für die mikrobielle Residualmasse (abgestorbene mikrobielle organische Substanz) fungieren. Denn Aminozucker sind im Boden überwiegend Bestandteile von Mikroorganismen (Pilze und Bakterien). Mikroorganismen spielen im Boden eine zentrale Rolle. Sie sind verantwortlich für die Bodenfruchtbarkeit und – qualität. So wandeln die Mikroorganismen organisches Material (Ernteteste oder organische Dünger) in pflanzenverfügbare Nährstoffe um (Balot et al., 2003, Wu et al., 1993). Weiterhin wächst im Kontext der globalen Erwärmung die Bedeutung der C-Sequestrierung (Humusanreicherung) von Böden (Bodé et al., 2009). Böden sind in der Lage schätzungsweise 2300 Gt Kohlenstoff zu sequestrieren (IPCC, 2007). Um jedoch fundierte Aussagen über sequenstrierten Kohlenstoff treffen zu können, ist ein besseres Verständnis der Kohlenstoffbiogeochemie unverzichtbar. Die Biogeochemie hängt wiederum von der mikrobiellen Gemeinschaft innerhalb eines Ökosystems ab (Amelung, 2001).

Im Boden steuern Aminozucker 5 bis 12% des Gesamtstickstoffs bei (Stevenson, 1982) und ungefähr 3% des organischen Bodenkohlenstoffs (Jörgensen und Meyer, 1990). Es gibt bis zu 26 Aminozucker (Sharon, 1965), wobei bisher nur Glucosamin, Galaktosamin, Muraminsäure und Mannosamin in Böden quantifiziert werden. Da höhere Pflanzen Aminozucker nicht in signifikanten Mengen enthalten, können sie als Indikatoren für die Akkumulation von mikrobiellen Residuen (abgestorbene organische mikrobielle Substanz) im Boden (Amelung et al., 2002; Amelung 2003; Liang et al. 2006, 2007a/b) und für die mikrobielle Biomasse (lebende organische mikrobielle Substanz) auf frisch besiedelten Substraten wie Weizenstreu herangezogen werden. So ist während des Abbaus von Pflanzenmaterial wie zum Beispiel von Buchenblattstreu die kontinuierliche Zunahme des Aminozuckergehaltes speziell nur auf die mikrobiellen Abbauprozesse zurückzuführen (Scholle et al., 1993). Im Gegensatz zu anderen Zellwandkomponenten wie Ergosterol oder Phospholipidfettsäuren, die nach dem Zelltod sehr

1. Einleitung

schnell abgebaut werden, haben Aminozucker die Eigenschaft zur Akkumulation in Huminstoffen (Bondietti et al., 1972) und anderen Komponenten des organischen Bodenmaterials (Jörgensen and Meyer, 1990, Coelho et al., 1997, Amelung, 2001). Somit können nur Aminozucker als Indikatoren der mikrobiellen Biomasse für die Bestimmung der mikrobiellen Besiedlung auf lebendem oder gerade abgestorbenem Pflanzenmaterial zu Rate gezogen werden. Weiterhin konnte wiederholt gezeigt werden, dass Aminozucker nützliche Indikatoren für die Akkumulation von verschiedenen Arten mikrobieller Residuen in Böden sind (Amelung et al., 1999, Guggenberger et al., 1999, Amelung et al., 2002, Amelung, 2003, Liang et al., 2007 ab).

Eine besondere Bedeutung kommt dabei den Aminozuckern Glucosamin und Muraminsäure zu. Diese Aminozucker sind spezifisch für zwei Hauptgruppen von Bodenmikroorganismen. So kommt Muraminsäure ausschließlich in der bakteriellen Zellwand insbesondere im Mureingerüst Gram-positiver Bakterien vor (Millar und Casida, 1970, Kenne und Lindburg, 1983). Glucosamin kommt hauptsächlich in der pilzlichen Zellwand vor (Appuhn und Jörgensen, 2006). Bakterien enthalten ebenfalls Glucosamin als Bestandteil des Zellwand-Peptidoglykans. Durch die von Appuhn und Jörgensen (2006) bestimmten Umrechnungsfaktoren ist es möglich den pilzlichen und den bakteriellen Anteil in mikrobiellen Residuen zu bestimmen. So wird pilzlicher Kohlenstoff durch Multiplikation des pilzlichen Glucosamins mit 9 und bakterieller Kohlenstoff durch Multiplikation mit 45 erhalten. Wobei pilzliches Glucosamin in Wurzel- und Bodenhydrolysaten durch das Subtrahieren des bakteriellen Glucosamins vom Gesamtglucosamin (mol Glucosamin – mol Muraminsäure) erhalten wird. Dabei wird angenommen, dass Muraminsäure und Glucosamin im Verhältnis 1 zu 2 in bakteriellen Zellwänden vorhanden ist (Engelking et al., 2007b). Die Funktion des theoretischen Verhältnisses von Glucosamin und Muraminsäure in der bakteriellen Zellwand von 1 zu 1 ist unbekannt (Zelles und Alef, 1995).

Wenig ist bekannt über den Ursprung von Galaktosamin im Boden, obwohl es einen Anteil von 30 bis 50% des Aminozuckergehalts ausmacht. Galaktosamin steuert durchschnittlich 4% in Bakterienkulturen und 15% in Pilzkulturen zum Gesamtaminozuckergehalt bei. Engelking et al. (2007b) konnten nach der Zugabe von Glucose zum Boden keine Veränderungen bezüglich des Galaktosamingehalts feststellen.

Noch weniger ist über den Ursprung von Mannosamin bekannt (Amelung 2001; Amelung et al. 2008), obwohl Autoren wie Amelung et al. (1999) und Guggenberger et al. (1999) signifikante Mengen an Mannosamin in verschiedenen Böden bestimmt haben. Mannosamin ist Bestandteil vieler Organismen. So kommt es im pilzlichen Melanin (Coelho et al. 1997) und als Komponente von Sialinsäure vor. Sialinsäure ist in Pilzen wie *Aspergillus fumigatus* (Wasylnka et al., 2011) und in Bakterien (Ferrero und Aparicio. 2010, Yoneyama et al., 1982) enthalten.

1.2 Aminozuckeranalytik – Allgemeine Chemie der Aminozucker

Für ein besseres Verständnis der Aminozuckeranalytik ist es wichtig die chemische Struktur und damit auch die chemischen Eigenschaften der zu bestimmenden Aminozucker zu kennen. Aminozucker sind Monosaccharide, deren Hydroxylgruppe durch eine Aminogruppe ersetzt ist (Abb.1). Dabei wird zwischen Glycosylaminen und Aminodesoxyzuckern unterschieden. Bei Glycosylaminen ist das Stickstoffatom an das anomere Kohlenstoffatom gebunden, bei Aminodesoxyzuckern ersetzt das Stickstoffatom das Sauerstoffatom an einer anderen Stelle im Molekül. Somit handelt es sich nur dann um Aminozucker, wenn das Stickstoffatom das Sauerstoffatom nicht am anomeren Kohlenstoffatom ersetzt hat (Vollhardt und Schore, 2000).

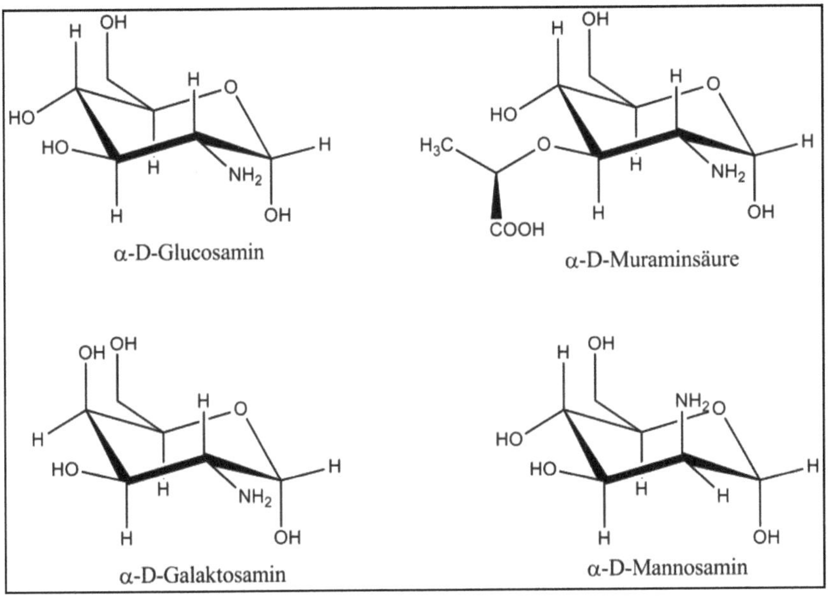

Abb. 1 Strukturformeln der Aminozucker-Monomere

In der Natur kommen Aminozucker nur gebunden vor. So sind im Chitin, einem Polysaccharid, die Acetylglucosamineinheiten β-1,4-glykosidisch gebunden (Abb.2).

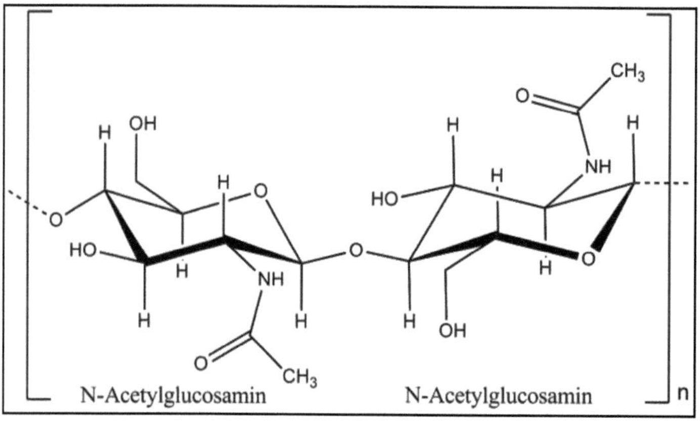

1. Einleitung

Abb. 2 Baueinheit von Chitin

Wohingegen im Peptidoglucan der bakteriellen Zellwand N-Acetylglucosamin und N-Acetylmuraminsäure β-1,4-glykosidisch miteinander verknüpft sind und lineare Kettenmoleküle bilden. Dabei ist an N-Acetylmuraminsäure eine Kette aus 4 Aminosäuren angeknüpft (Abb.3). Bei grampositiven Bakterien ist in dem Tetrapeptid Diaminopimelinsäure durch L-Lysin ersetzt.

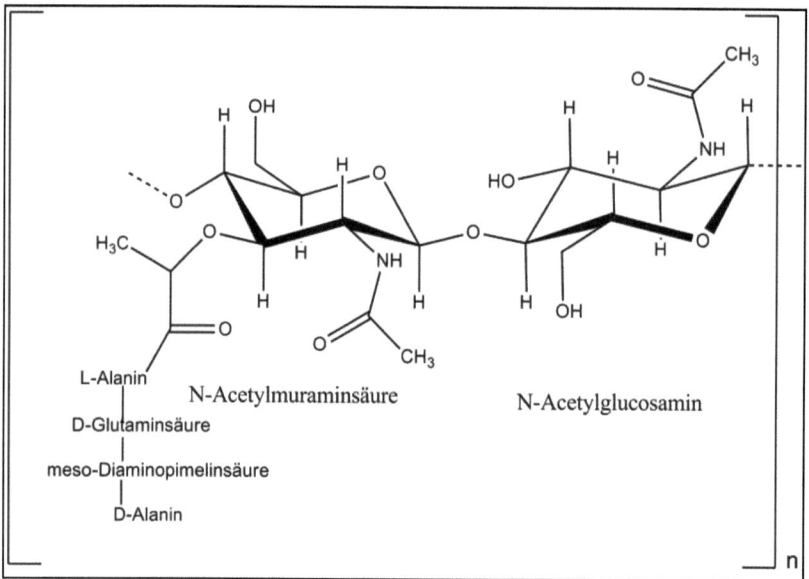

Abb. 3 Aufbau des Peptidoglycans (Murein) gramnegativer Bakterien

1. Einleitung

Um Aminozucker quantitativ bestimmen zu können, ist es notwendig, dass sie als Monomere vorliegen. Aminozucker sind im Gegensatz zu Zuckern aufgrund der Aminogruppe sehr stabil. Somit kann eine Hydrolyse, welche die Polymere in Monomere spaltet, mit Salzsäure erfolgen. Bei Verwendung von starken Mineralsäuren wie Salzsäure ist eine nahezu komplette Freisetzung der Monomere aus ihrem Polymerverband gewährleistet.

Durch ihre Aminogruppe zeigen Aminozucker ein basisches Verhalten, indem sie im sauren und neutralen Milieu ein Proton an das freie Elektronenpaar des Stickstoffs anlagern. Ein davon abweichendes chemisches Verhalten ist bei der Muraminsäure zu finden. Aufgrund ihrer Carboxylgruppe reagiert Muraminsäure nicht nur basisch sondern auch sauer. Ähnlich wie bei Aminosäuren (Vollhardt und Schore, 2000) ist davon auszugehen, dass Muraminsäure im sehr sauren Bereich (pH < 1) als diprotoniertes Kation (RNH_3^+ und R-COOH), im neutralen Bereich (pH 6-7) als monoprotoniertes Zwitterion (RNH_3^+ und $R-COO^-$) und im basischenBereich (pH>13) als deprotoniertes Aminocarboxylat-Ion vorliegt (RNH_2 und $R-COO^-$). , wobei R den Rest des Moleküls darstellt. Dieses abweichende Verhalten der Muraminsäure von den anderen Aminozuckern erschwert die Analytik der underivatisierten Aminozucker eminent.

1.3 Aminozuckeranalytik – Chromatographie der Aminozucker

In underivatisierter Form liegen die basischen Aminozucker (Glucosamin, Galaktosamin und Mannosamin) im neutralen und sauren Milieu (pH 2-7) in protonierter Form vor, wohingegen Muraminsäure monoprotoniert vorliegt. Für die Trennung von Kationen ist in der Flüssigkeitchromatographie (LC) die Kationenaustauschchromatographie vielversprechend. Bei der Kationenaustauschchromatographie sind an der stationären Phase Anionen (Abb.4a) (hier: Sulfonsäureionen) gebunden, die aufgrund der negativen Ladung positiv geladene Teilchen binden und retardieren können (Cammann, 2001). Die Aminozucker werden mit einem schwach sauren gepufferten Eluenten auf die Säule befördert und liegen dabei in positiv geladener Form vor. Eine Ausnahme bildet hier jedoch Muraminsäure. Die negativ geladene Carboxylgruppe der Muraminsäure verkürzt die Retardierung auf dem Säulenmaterial. Somit reagiert Muraminsäure nur sehr gering mit dem Säulenmaterial und eluiert kurz nach der Totzeit des Systems. Eine Trennung erfolgt aufgrund der unterschiedlichen Affinitäten der Aminozucker zu dem Säulenmaterial. Im basischen Milieu liegen die Aminozucker deprotoniert vor. Hier erfolgt für die basischen Aminozucker, ähnlich wie bei Einfachzuckern, eine Bindung mit dem positiv geladenen Harz und den negativ geladenen Hydroxylgruppen (Abb.4b). Da die Carboxylgruppe der Muraminsäure eine viel stärkere Affinität zu dem Säulenmaterial hat als die Hydroxylgruppen des basischen Eluenten

(Natriumhydroxid), verbleibt sie länger auf der Säule und kann nur durch Eluenten mit hoher Elutionskraft (z.B. Acetat) eluiert werden

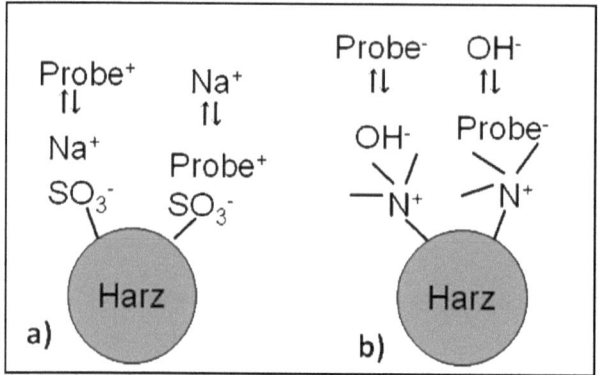

Abb. 4 a) Prinzip des Trennmechanismus Kationenaustauschchromatographie, **b)** Anionenaustauschchromatographie

Nach erfolgreicher Trennung werden die Aminozucker detektiert. Dies erfolgt entweder mittels Nachsäulenderivatisierung und Fluoreszenzdetektion oder mittels massenspektrometrischen Detektoren. Die chromatographische Trennung mittels Anionen- oder Kationenaustauschchromatographie kann auch zur Probenaufbereitung verwendet werden. So können z.B. bei Anwendung der Kationenaustauschchromatographie die Aminozucker von neutralen und negativ geladenen Verunreinigungen aus der Matrix abgetrennt werden.

Werden die Aminozucker jedoch vor der chromatographischen Trennung zu neutralen und hydrophoben Molekülen (fluoreszierende Isoindolderivate) mittels Ortho-phthaldialdehyd (OPA) in fluoreszierende Isoindolderivate derivatisiert (Abb.5), ist eine Trennung durch die Umkehrphasenchromatographie (Abb.6a) sehr geeignet.

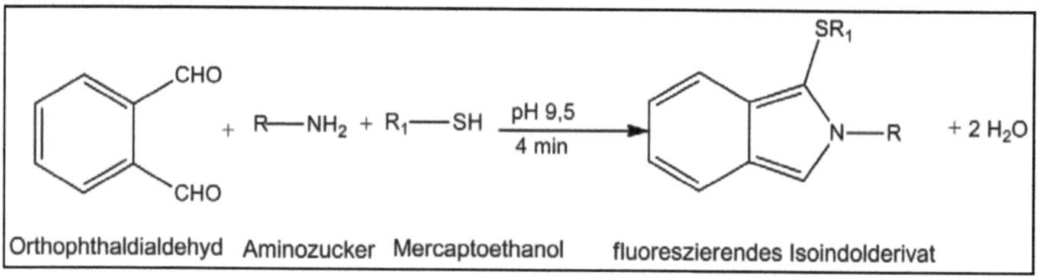

Abb. 5 Reaktionsmechanismus der OPA- Derivatisierung

1. Einleitung

Die Trennung erfolgt dabei an einer Umkehrphase (RP-Phase) (hydrophobe feste stationäre Phase mit C18-Ketten und aktiven Silanolgruppen; hier:Hypersil Octadecylsilan (ODS) – Säule) (Abb.6b), wobei der Analyt Wechselwirkungen mit der stationären Phase eingeht (Abb. 6a). Die Wechselwirkungen sind in erster Linie polare Wechselwirkungen wie Dipol-Dipol-Reaktionen und Wasserstoffbrückenbindungen) sowie sterische Aspekte. Weiterhin spielen hydrophobe Wechselwirkungen eine mögliche jedoch untergeordnete Rolle für die Trennung an einer RP-Phase (Cammann, 2001). Bei dieser Methode übt die Derivatisierung der Analyten mit OPA auch Einfluss auf die Güte der Trennung aus. Durch die Derivatisierung der Aminozucker mit OPA werden fluoreszierende, hydrophobe, neutrale Isoindolderivate erhalten, die aufgrund ihrer Molekülstruktur zu Wechselwirkungen basierend auf sterischen Aspekten mit dem Säulenmaterial befähigt sind. Demzufolge ist für diese Methode eine Nachsäulenderivatisierung eher ungeeignet

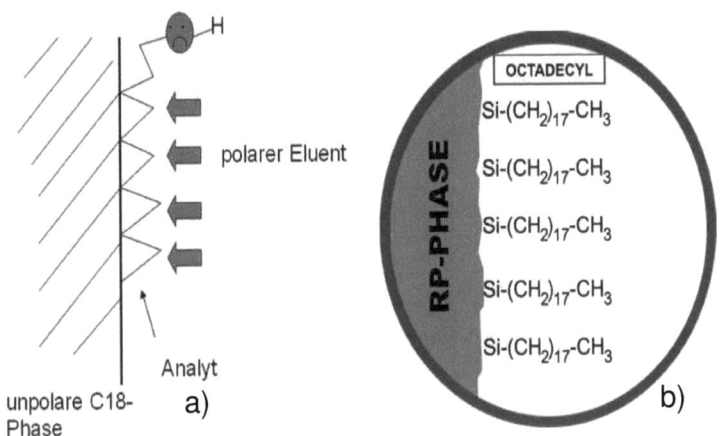

Abb. 6 a) Prinzip des Trennmechanismus Umkehrphasen(RP)-Chromatographie, **b)** Säulenmaterial Umkehrphasen(RP)-Chromatographie

1.4 Aminozuckeranalytik – Isotopenmassenspektrometrie

Die Bestimmung stabiler Isotope ist ein wichtiger Bestandteil innerhalb der Umweltforschung und gewinnt zunehmend an Bedeutung für die Grundlagenforschung in Bereichen wie Lebensmittelchemie, Biochemie sowie Klinische- und Pharmazeutische Chemie (Monsadl et al., 2007). Kohlenstoff hat drei natürlich vorkommende Isotope (^{12}C, ^{13}C, und ^{14}C), wobei ^{12}C und ^{13}C stabile und ^{14}C radioaktive Isotope sind. Die natürliche Abundanz in der Umwelt beträgt für ^{12}C ca. 98,89%, für ^{13}C 1,11% (Boutton, 1996) und für ^{14}C $<10^{-10}$ % (Goh, 1991). Die unterschiedliche Anzahl an Neutronen im Kern stabiler Isotope eines Elements haben Auswirkungen auf

1. Einleitung

physikalische und chemische Eigenschaften von Molekülen, die diese Isotope enthalten. Diese sogenannten Isotopeneffekte machen sich in der Kinetik biochemischer Effekte, welche an der Entstehung und am Abbau organischer Substanz beteiligt sind, bemerkbar. So führt beispielsweise die Diskriminierung des reaktionsträgeren Kohlenstoffisotops (^{13}C) zu einer Abreicherung in gebildeter organischer Substanz im Vergleich zu dem atmosphärischen CO_2. Diese Abreicherung bzw. Defizit an CO_2 ist nicht für alle Pflanzen gleich, sondern vom Photosynthesemechanismus der Pflanzen abhängig. Bei C3 Pflanzen stellt das erste fassbare Produkt der CO_2-Fixierung eine C3-Verbindung (3-Phosphoglycerinsäure) dar. Dabei wird das CO_2 an die Ribulose-1,5-Bisphosphat gebunden, wobei ein Zwischenprodukt mit sechs C-Atomen entsteht (C6-Körper), das sofort in zwei C3-Körper (PGS =3-Phosphoglycerinsäure) zerfällt. Diese Reaktion wird durch das Enzym Ribulose-Biphosphat-Carboxylase/Oxygenase (RuBisCo) katalysiert. Bei C4-Pflanzen ist das erste fassbare Produkt der CO_2-Fixierung eine C4-Verbindung (Oxalessigsäure). Dabei wird CO_2 durch die Phosphoenolpyruvat (PEP)-Carboxylase in das Phosphenolpyruvat eingebaut, wodurch nach Abspaltung des anorganischen Phosphats Oxalessigsäure entsteht (Nultsch, 2000). So diskriminieren C4 Pflanzen weniger gegen $^{13}CO_2$ als C3 Pflanzen und haben daher auch einen höheren delta-Wert (Abb.7). Der delta (δ)- Wert ist ein häufig verwendetes Maßsystem zur Angabe von Isotopenverhältnissen. Er beschreibt die Abweichung des Isotopenverhältnisses relativ zu einem Standard. Der Standard zur Bestimmung des $\delta^{13}C$- Wertes ist auf den international anerkannten PDB (PeeDeeBelmnite)-Standard festgelegt. PDB ist ein fossiler Kohlenstoff aus der Kreidezeit *Bellemnitella americana* aus den Peedee Formation in South Carolina, USA (Carle, 1991). Aufgrund der unterschiedlichen Photosynthesemechanismen liegen die $\delta^{13}C$-Werte für C4-Pflanzen zwischen -9 und -17‰ PDB und für C3-Pflanzen zwischen -22 und -32 δ PDB (Boutton, 1996). So können, das System Boden-Pflanze betrachtend, bei einer Vegetationsänderung (C3 zu C4-Pflanzen) Veränderungen im C4-bürtigen Kohlenstoff im Boden, in der mikrobiellen Biomasse sowie im CO_2 des Bodens gemessen werden (Gregorich et al., 1996, Collins et al., 1999, Ludwig et al., 2003).

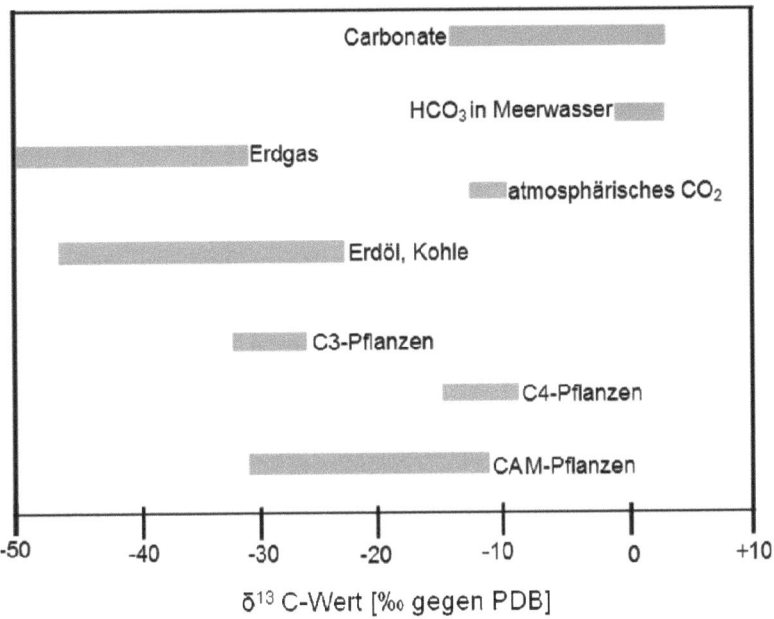

Abb. 7 ^{13}C-Isotopengehalt der wichtigsten Kohlenstoff reservoire der Erde (Carle, 1991)

Auch Mikroorganismen sind in der Lage die δ^{13}C zu diskriminieren. In einem ungestörten Boden-Pflanze- System hat die organische Substanz des Bodens ungefähr den gleichen δ^{13}C-Wert wie die Pflanzengesellschaft (Boutton, 1996). Die mikrobielle Biomasse nimmt die in Boden eingearbeiteten Materialien auf und wandelt sie in neue Komponenten um. Bei dieser Umwandlung kommt es zu Isotopeneffekten. Zum Beispiel veratmen Mikroorganismen CO_2 mit geringerem ^{13}C-Anteil, wohingegen schwere Moleküle in der mikrobiellen Biomasse und der restlichen organischen Bodensubstanz angereichert werden (Nadelhofer und Frey, 1988). Andere Autoren wie Agren et al. (1996) gehen aber davon aus, dass bei einer Mineralisation von komplexen Komponenten (z.B. Lignin) mit hohem ^{12}C-Anteil die verbleibende organische Substanz ^{13}C angereichert ist, während bei Verbrauch von leicht verfügbaren Komponenten, die ^{13}C angereichert sind, das verbleibende Material ^{13}C abgereichert ist.

Mit der Isotopenverhältnismassenspektrometrie (IRMS) ist es möglich diese Effekte zu bestimmen. Die Kopplung eines Isotopenverhältnismassenspektrometers mit einem Hochleistungsflüssigkeitschromatographie (HPLC) - System bietet die Möglichkeit die Isotopie einzelner Analyte zu bestimmen. Dadurch wird es möglich die Umsatzdynamiken bakterieller und pilzlicher Mikroorganismen direkt zu erfassen.

2. Ziele der Arbeit

Ziel der Arbeit ist es, eine aminozucker-spezifische $\delta^{13}C$ Methode mit hoher Genauigkeit zur Erfassung der Bildung und Nutzung von mikrobiellen Residuen in Böden- und Pflanzenmaterialien zu etablieren und anzuwenden. Die von Glaser und Gross (2005) beschriebene aminozucker-spezifische $\delta\ ^{13}C$-Methode mittels Gaschromatographie-verbrennungs (GC-C)-IRMS ist aufgrund der vorherigen offline-Derivatisierung der Aminozucker zeitaufwendig und ungenau. Die Ungenauigkeit der Methode ist auf die Derivatisierung zurückzuführen, da diese zu Fraktionierung und zur Einbringung zusätzlicher C-Atome führt. Dafür soll zunächst die HPLC-RP-Methode auf der Basis von (Appuhn et al., 2004) modifiziert und an die verfügbare HPLC adaptiert werden. Nach erfolgreicher Validierung dieser HPLC-RP-Methode ist eine weitere Methode zur Bestimmung von Aminozuckern an einer Hochleistungsanionenaustauschchromatographie (HPAEC) (Bodé et al., 2009) zu optimieren und zu validieren. Die Einführung der HPAEC-Methodik ist erforderlich, um kohlenstofffreie Messungen durchzuführen, die wiederum für die spätere Messung der $^{13}C/^{12}C$- Verhältnisse mit einem IRMS notwendig sind. Nach der Kalibrierung der HPAEC-Methode soll diese mit der RP-HPLC-Methode verglichen werden. Weiterhin sind andere chromatographische Methoden mit kohlenstofffreien Eluenten zu testen (Kationenaustausch- und Ionenausschlusschromatographie) in Hinblick auf eine schnellere und effizientere Trennung der Aminozucker. Die verschiedenen HPLC-Methoden sollen dann mit der validierten RP-HPLC-Methode verglichen werden.

Zusätzlich sind verschiedene Purifikationsmethoden zur Eliminierung von Matrixsubstanzen und zur Konzentrierung von Muraminsäure in den Probenhydrolysaten zu testen. Die Konzentrierung von Muraminsäure ist notwendig, da die LC-IRMS im Vergleich zur RP-HPLC eine um ca. 400 fach geringere Empfindlichkeit bezüglich Muraminsäure hat. Als Purifikationsmethoden können unter anderem Kationen- und Anionenaustauscherharze verwendet werden.

Nach erfolgreicher Kalibrierung der HPAEC-IRMS wird mit der neu eingeführten Methodik ein Pilzwachstumsversuch durchgeführt, um die Bildung und den Umsatz saprotropher Pilze zu bestimmen. Der Hintergrund des Pilzwachstumsversuchs sind folgende Fragestellungen:

a) Welches Substrat wird bevorzugt von den saprotrophen holzzersetzenden Pilzen abgebaut?

b) Worauf lässt sich die in der Literatur beschriebene $\delta^{13}C$-Anreicherung in den saprotrophen Pilzen während des Pilzwachstums zurückführen? Inwiefern spielen kinetische Isotopenfraktionierung und Inkorporation des Substrat-Kohlenstoffs dabei eine Rolle?

3. Optimisation of amino sugar quantification by HPLC in soil and plant hydrolysates

Biology and Fertility of Soils

Caroline Indorf [1]*, Jens Dyckmans [2], Khalid S. Khan [1]*, Rainer Georg Joergensen [1]

[1] Department of Soil Biology and Plant Nutrition, University of Kassel Nordbahnhofstr. 1a, 37213 Witzenhausen, Germany

[2] Centre for Stable Isotope Research and Analysis, University of Göttingen, Büsgenweg 2, 37077 Göttingen, Germany

Abstract

Amino sugars are increasingly used as indicators for the accumulation of microbial residues in soil and plant material. A reverse-phase high-performance liquid chromatography (HPLC) method was improved for the simultaneous determination of muramic acid, mannosamine, glucosamine, and galactosamine in soil and plant hydrolysates via ortho-phthaldialdehyde (OPA) pre-column derivatisation and fluorescence detection. The retention time was reduced and the separation of muramic acid and mannosamine was optimised by modifying the mobile phase. The effects of excitation wavelength, OPA reaction time, tetrahydrofuran concentration and pH value of the mobile phase on the amino sugar separation were tested. Quantification limits were in the range of 0.13 to 0.90 µg ml^{-1}. No interferences exist from amino acids or other primary amines, occurring in soil and plant hydrolysates.

Key words: amino sugars / HPLC / ortho-phthaldialdehyde / microbial residues

---

* Corresponding author. Tel.: + 49 5542 98 1503; e-mail: cindorf@uni-kassel.de

** Permanent address: Department of Soil Science, PMAS-Arid Agriculture University, Murree Road, 46300 Rawalpindi, Pakistan

3. Optimisation of amino sugar quantification by HPLC in soil and plant hydrolysates

3.1. Introduction

Amino sugars make a significant contribution of 5 to 12% to soil organic N (Stevenson 1982) and roughly 3% to soil organic C (Joergensen and Meyer 1990). Up to 26 amino sugars have been found in microorganisms (Sharon 1965), whereas four of them have been quantified in soil. These amino sugars are glucosamine, galactosamine, muramic acid and mannosamine (Amelung et al. 2008). It has repeatedly been shown that amino sugars are useful indicators for the accumulation of different types of microbial residues in soil (Amelung et al. 2002; Amelung 2003; Liang et al. 2006, 2007a/b). Fungal cell walls are the major source of glucosamine in soils (Appuhn and Joergensen 2006). Muramic acid occurs exclusively in bacterial cell walls, especially in the murein skeleton of Gram-positive species (Millar and Casida 1970; Kenne and Lindburg 1983). Also bacteria contain glucosamine in their peptidoglycan cell wall, but only the glucosamine that occurs in excess to muramic acid may be attributed to fungal sources (Chantigny et al. 1997; Guggenberger et al. 1999; Amelung 2001). Appuhn and Joergensen (2006) determined average conversion factors of 9 to convert fungal glucosamine to fungal C and 45 to convert muramic acid to bacterial C.

Galactosamine contributes roughly one third to the total sum of amino sugars observed in soil and is also nearly exclusively of microbial origin (Engelking et al. 2007). Galactosamine contributed on average 4% to the total amino sugar concentration in cultured bacteria and 15% in cultured fungi (Engelking et al. 2007). However, the function of galactosamine within bacterial and fungal cells and consequently the processes behind galactosamine formation during decomposition processes are still unknown. However, galactosamine in soil is still attributed to be mainly of bacterial origin (Amelung et al. 2008), which has been sometimes supported by correlation analysis (Rottmann et al. 2010). Virtually nothing is known about the origin of mannosamine (Amelung 2001; Amelung et al. 2008), although Amelung et al. (1999) and Guggenberger et al. (1999) found significant amounts in different soils. Mannosamine has been found in fungal melanin (Coelho et al. 1997) and as component of sialic acids of *Aspergillus fumigatus* on their conidial surface (Wasylnka et al. 2001). Mannosamine, i.e. N-acetyl D-mannosamine containing sialic acids are present as protective capsular components of bacteria, invading mammals (Ferrero and Aparicio 2010). However, small amounts of mannosamine may be also present as common linkage units between peptidoglycan and other bacterial cell wall components such as glycerol teichoic acid (Yoneyama et al. 1982).

Several methods for the determination of amino sugars have been published. Highly specific gas chromatographic analyses require difficult off-line derivatisation steps, e.g. derivatisation of the hydrolysis products in volatile aldononitrile acetates (Zhang and Amelung 1996). High performance liquid chromatography (HPLC) methods need either derivatisation steps (Diaz et al. 1996; Ekblad and Näsholm 1996; Appuhn et al. 2004) or special equipment for anion exchange chromatography combined with pulsed amperometric detection (Kaiser and Benner 2000; Benner and Kaiser 2003). This means that HPLC methods are generally less time consuming as they employ no or automated on-line derivatisation.

Appuhn et al. (2004) were the first to describe a method to determine the four amino sugars simultaneously. However, this methodological approach was hampered by two drawbacks: (1) The method was reliably working only on the Agilent 1100 HPLC equipment and (2) mannosamine was insufficiently separated in most soil hydrolysates, which resulted in erroneous high values usually omitted in further publications (Appuhn and Joergensen 2006; Engelking et al. 2007). The objective of the present paper was improving the method of Appuhn et al. (2004) to give reliable results for all four amino sugars in soil and plant hydrolysates using different HPLC equipments.

3.2. Materials and methods

3.2.1. Soil and plant samples

Method optimisation was implemented using six different soil samples (0-10 cm) taken from four arable and two forest sites in Germany (Hessia and Lower Saxony) differing in physical, chemical and microbial properties (Table 1) and five different plant litter samples (Table 2). Soil physical, chemical and biological properties of the soil samples and total C and N of litter samples were determined as described by Probst et al. (2008).

3. Optimisation of amino sugar quantification by HPLC in soil and plant hydrolysates

Table 1 Physical, chemical and microbiological properties of the soils used in this investigation

Soil	Clay	Silt	Sand	pH-H_2O	Soil organic C	Total N	Microbial biomass C	Ergosterol
	%				mg g^{-1} soil		µg g^{-1} soil	
Forest 1	15	77	8	3.9	59.9	4.3	610	3.2
Forest 2	6	39	55	3.9	58.0	2.4	520	5.4
Arable 1	35	55	10	7.0	18.6	1.8	450	1.6
Arable 2	34	56	10	7.3	15.1	1.4	360	1.1
Arable 3	18	66	16	7.8	14.0	1.2	200	0.5
Arable 4	8	8	84	7.4	7.8	0.7	180	0.6

Table 2 Chemical properties of the plant litter materials used in this investigation

Litter	Total C	Total N
	µg g^{-1} dry weight	
Sugarcane filter cake	448	31.4
Maize leaves	440	22.0
Pea leaves	443	14.3
Amaranth straw	397	8.2
Wheat straw	440	4.1

The amino sugar extraction was based on the method described by Appuhn et al. (2004) with minor modifications. Sieved (< 2 mm) and air-dried soil (400 mg) or oven-dried (40 °C) and steel ball-milled (Retsch, Haan, Germany) plant material (700 mg) was mixed with 10 ml of 6 M HCl. After 6 h (soil) or 3 h (plant material) hydrolysis at 105°C the samples were filtered over glass filters (Whatman GF/A). For the determination of the recovery rate, 0.3 ml of a 150 µmol l^{-1} standard solution was added to 0.3 ml of a quartz sand hydrolysate. A 0.3 ml aliquot was evaporated to dryness at 40-45 °C to remove HCl, re-dissolved in water, evaporated a second time and re-dissolved in 1 ml water. To test the effect of sample pH on the amino sugar amounts another set of samples was prepared using 1 ml phosphate buffer solution pH 7 (containing 40.8 mmol l^{-1} $Na_2HPO_4 \times 2H_2O$ and 25.9 mmol l^{-1} KH_2PO_4) to re-dissolve samples. After centrifugation at 5000 g, the supernatant was frozen and stored at -18 °C until analysis.

3.2.2. Solutions

All solutions were prepared with Milli-Q water produced via a Direct-Q 3 system (Millipore, Billerica, MA, USA). All other reagents were of high purity ($\geq$ 95 %). The buffer solution (pH 11) was prepared by dissolving 50 g H_3BO_3 in 900 ml water, adjusted to pH 11 with KOH (47% solution) and diluted to 1 l with water. This solution was stable for up to 12 months at 4 °C. The reducing solution was prepared by adding 2.5 ml 2-mercaptoethanol to 100 ml buffer solution. This solution was stable for up to 6 months at 4 °C in the dark. The derivatisation reagent was prepared by dissolving 25 mg ortho-phthaldialdehyde (OPA) in 2 ml methanol, mixed with 2 ml of reducing solution and diluting to 44 ml with buffer solution. The reagent was stable for up to 7 days at 4 °C in the dark.

For the standard stock solutions of the four amino sugars, standards (Sigma Aldrich, St. Louis, MO, USA) were dissolved in water to a concentration of 1000 µmol l^{-1} (mannosamine, galactosamine, glucosamine) and 100 µmol l^{-1} (muramic acid) respectively and stored at –18°C. Standard working solutions were prepared by diluting 4 aliquots of the stock solutions to a concentration range between 210 µmol l^{-1} and 5 µmol l^{-1} (mannosamine, galactosamine, glucosamine) and 21 µmol l^{-1} and 0.5 µmol l^{-1} (muramic acid), respectively. The working standard solutions were stable for over 12 months at -18 °C.

For evaluating the interference of common amino acids and related compounds which are abundant in soil and plant hydrolysates, we measured two different amino acids standards (Sigma Aldrich, St. Louis, MO, USA). Amino acid standard solution 1 contained the following components at a concentration of 25 µmol l^{-1} and 12 µmol l^{-1} for L-cysteine, respectively: L-alanine, ammonium chloride, L-arginine, L-aspartic acid, L-cystine, L-glutamic acid, glycine, L-histidine, L-isoleucine, L-leucine, L-lysine, L-methionine, L-phenylalanine, L-proline, L-serine, L-threonine, L-thyrosin, L-valine. Amino acid standard solution 2 contained the following components at a concentration of 50 µmol l^{-1}: γ-amino-n-butyric acid, ammonium chloride, L-anserine, L-arginine, L-carnosine, creatinine, ethanolamine, L-histidine, L-homocystine, δ-DL-hydroxylysine, L-lysine, 1-methyl-L-histidine, 3-methyl-L-histidine, L-ornithine and L-tryptophan.

3.2.3 Chromatographic conditions

Chromatographic separations were performed on a Phenomenex (Aschaffenburg, Germany) Hyperclone C_{18} (ODS) column (125 mm length × 4 mm diameter, 5 µm particle size, 12 nm pore size), protected by a Phenomenex C_{18} security guard cartridge (4 mm length × 2 mm diameter). The column was placed in a column oven set at 35 °C. The HPLC system consisted of a Dionex (Germering, Germany) P 580 gradient pump, a Dionex Ultimate WPS – 3000TSL analytical autosampler with in-line split-loop injection and thermostat and a Dionex RF 2000 fluorescence detector set at 445 nm emission and 330 nm excitation wavelength with medium sensitivity.

An autosampler designed for automated pre-column derivatisation was used because of the need of a high injection precision and an effective external needle wash for eliminating carryover. Another autosampler tested (Dionex ASI 100) yielded poor reproducibility. Vials with OPA reagent and samples as well as vials for preparation were stored in the autosampler at 15 °C. For derivatisation, 50 µl OPA reagent and 30 µl sample were mixed in a preparation vial and after 120 sec reaction time 15 µl of the indole derivates were injected.

The mobile phase consisted of two eluents. Eluent A was a 97.8/0.7/1.5 (v/v/v) mixture of a water phase, methanol and tetrahydrofuran (THF). The water phase contained 52 mmol sodium citrate and 4 mmol sodium acetate, adjusted to pH 5.3 with HCl. Then methanol and THF were added. Eluent B consisted of 50% water and 50% methanol (v/v). For degassing and sterilisation both eluents were filtered over 0.2 µm pore size hydrophilic propylene membrane filters. The mobile phase was delivered at a flow rate of 1.5 ml min^{-1}. The amino sugar separation was performed isocratically but a gradient was used for cleaning the column after every run. Every run was starting at an eluent A/B v/v composition of 93/7 for 19 min. A linear gradient was run to reach 80% B after 3 min and remaining isocratic for 3 min. A reverse gradient to 93/7 within 3 min was followed by 2 min isocratic run after which the column is preconditioned for the next sample. It is important to start the OPA derivatisation only at the beginning of every run and not during column preconditioning of the previous run. Otherwise, the equilibration time is not long enough for a good reproducibility.

3.3. Results

3.3.1 Method optimisation

The chromatographic conditions used in the optimised method provided a good separation within approximately 19 min (Fig. 1). Retention times of examined amino sugars were: (1) muramic acid t_R = 8.4 min, (2) mannosamine t_R = 9.4 min, (3) galactosamine t_R = 15.4 min and (4) Glucosamine t_R = 16.6 min.

3. Optimisation of amino sugar quantification by HPLC in soil and plant hydrolysates

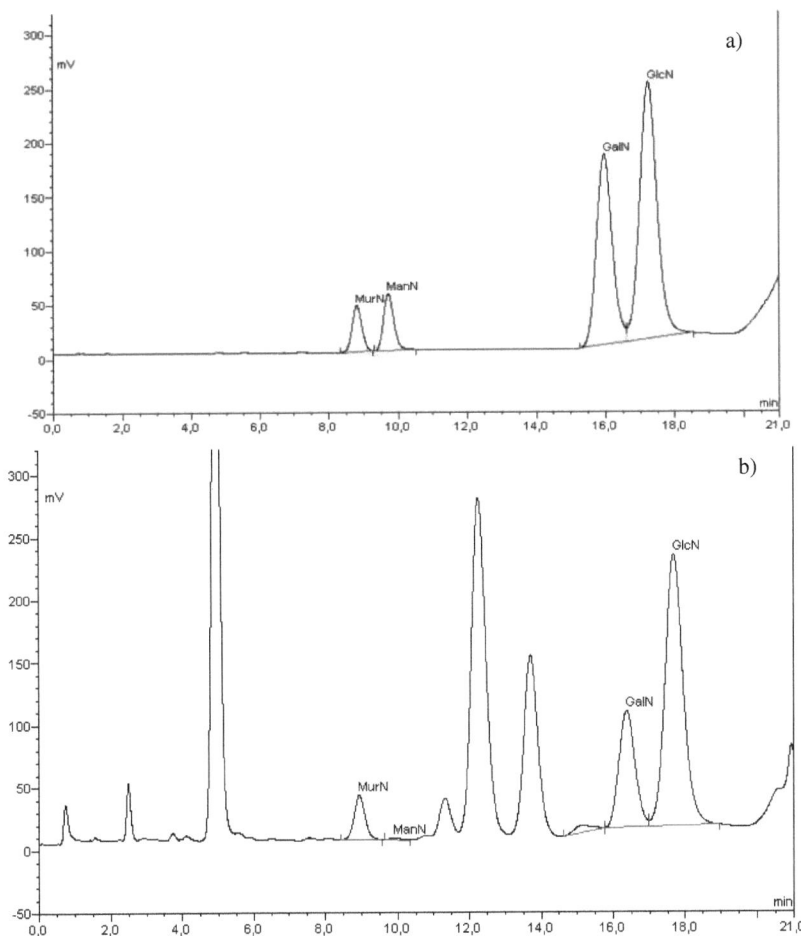

Fig.1 Chromatograms of (a) a standard mixture consisting of 13µmol l-1 muramic acid, 130 µmol l-1 mannosamine, 130 µmol l-1 galactosamine, 130 µmol l-1 glucosamine and of (b) a soil hydrolysate (arable 1).

The components of the amino acid standard solution 1 occurred at 4.7 min, 10.5 min, 11.7 min, 12.6 min and > 20.0 min, respectively (Fig. 2b). The retention times of the amino acid standard solution 2 were at 10.5 min, 11.8 min, 18.1 min and > 20.min, respectively (Fig. 2c).

3. Optimisation of amino sugar quantification by HPLC in soil and plant hydrolysates

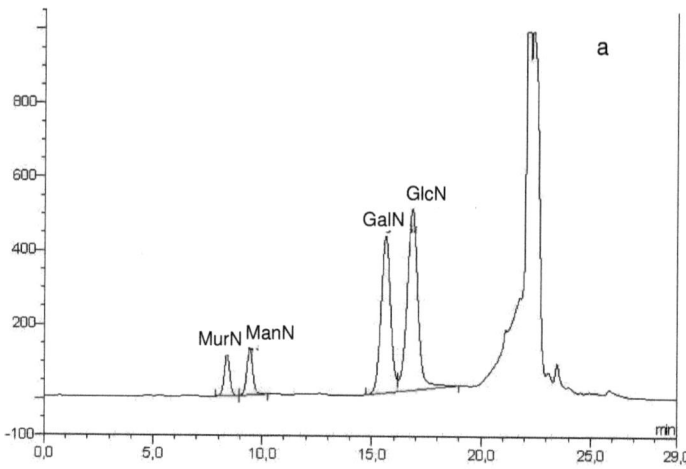

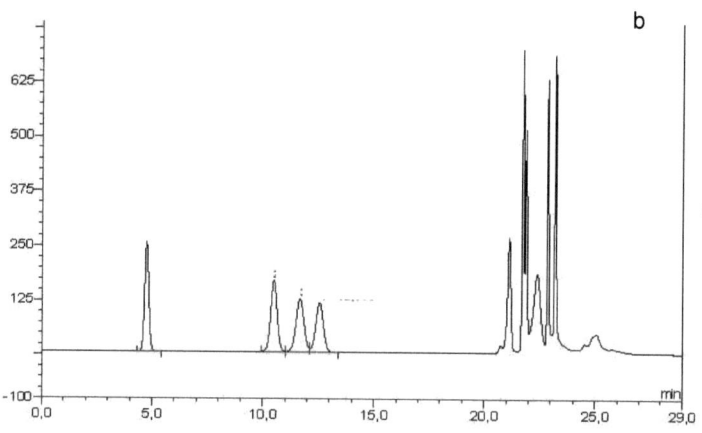

Fig. 2 Chromatograms of (a) an amino sugar standard mixture consisting of 17 µmol l^{-1} muramic acid, 170 µmol l^{-1} mannosamine, 170 µmol l^{-1} galactosamine, 170 µmol l^{-1} glucosamine and of (b) an amino acid standard 1.

3. Optimisation of amino sugar quantification by HPLC in soil and plant hydrolysates

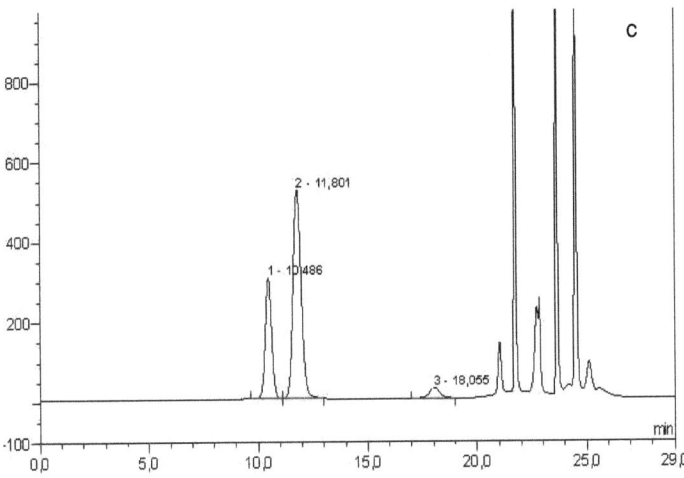

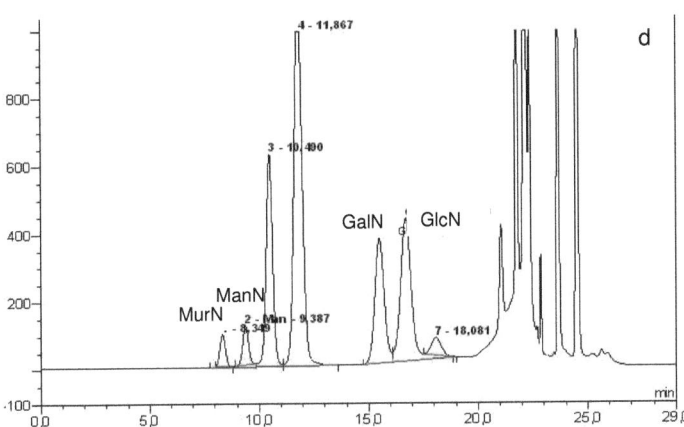

Fig. 2 Chromatograms of (c) an amino acid standard 2 and (d) an amino sugar standard spiked with an amino acid standard 2, components of amino acid standards (see materials and methods).

For testing the absence of any interference potential, an amino sugar standard (170 µmol l^{-1}) was mixed with the amino acid standard 2 (500 µmol l^{-1}) at the ratio of 5:1. All four amino sugar peaks were well separated from those of the amino acids (Fig. 2d).

Analysis of a standard mixture at different excitation wavelengths, different OPA reaction times and different pH values of the eluent revealed maxima for the area of all four amino sugar peaks at an excitation wavelength of 330 nm, an OPA reaction time of 120 sec an eluent pH of 5.3 (with the exception of mannosamine) respectively (Fig. 3). Increasing THF concentration in the eluent from 0.75% to 1.50% yielded a better resolution between muramic acid and mannosamine as well as to shorter retention times for all four amino sugars (Fig. 3d).

The calibration curves (Fig. 4) were linear in the range from 5 to 210 µmol l^{-1} (mannosamine, galactosamine, glucosamine) and from 0.5 to 21 µmol l^{-1} (muramic acid), respectively, with good correlation coefficients and standard deviations (Table 3). The coefficient of variation was roughly 2% for intraday and 5% for interday precision, respectively (Table 3). This reflects the reproducibility and the precision of the method excluding sample preparation. The accuracy (expressed as recovery) for the four analytes was determined by spiked quartz sand hydrolysate with the standard mixture solution. The results of the recovery of mannosamine, galactosamine and muramic acid ranged from 105 to 115%. The recovery rate for glucosamine was 81%. The limit of quantification (LOQ) depended on the amino sugar and varied from 0.5 to 5 µmol l^{-1} for the standard mixture solution (Table 3).

The content of the different amino sugars increased in the order mannosamine < muramic acid < galactosamine < glucosamine (Table 4). In three arable soils mannosamine was below the LOQ value. The glucosamine content in the soil samples ranged from 980 to 2730 µg g^{-1} soil and contributed between 55 to 62% to the total amino sugar content. Galactosamine and muramic acid contributed on average 35% and 5%, respectively, mannosamine, if present, only 2%. The glucosamine content in the plant litter material ranged from 36 (maize leaves) to 1550 µg g^{-1} dry weight (amaranth straw) and contributed 58 (maize leaves) to 93% (wheat straw) to the total amino sugar content (Table 5). Muramic acid was below the LOQ values in green pea and maize leaves. Mannosamine was below LOQ in maize leaves and wheat straw, but contributed roughly 20% to the total amino sugar content in pea leaves. Soil samples dissolved in water had significantly higher amino sugar contents than those dissolved in buffer solution (Table 4). In contrast, the sample solvent had no effect on the amino sugar amount in plant material (Table 5).

3. Optimisation of amino sugar quantification by HPLC in soil and plant hydrolysates

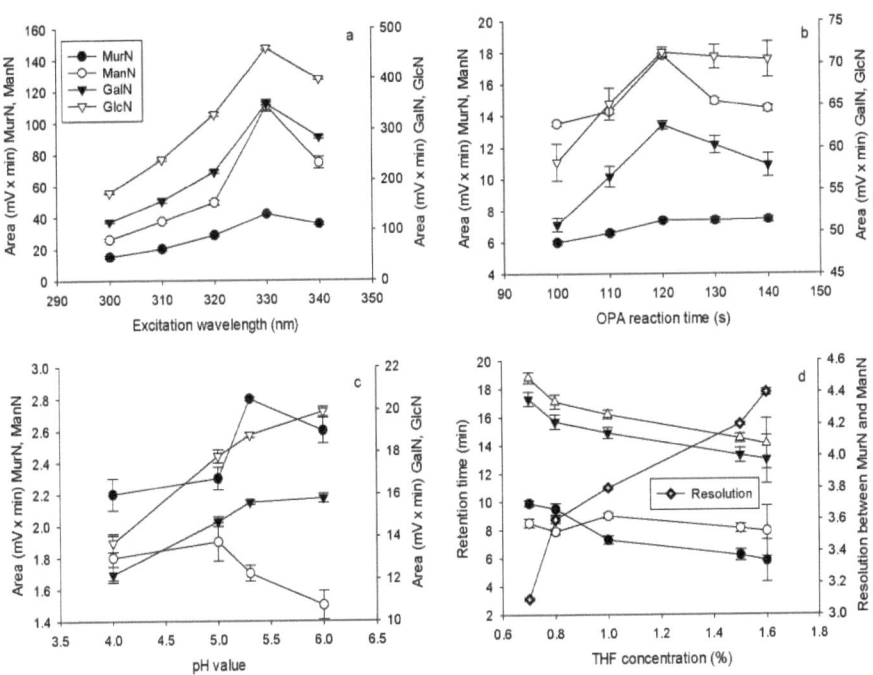

Fig. 3 Optimisation of HPLC parameters: (a) sensitivity of the amino sugar determination as a function of OPA reaction time; (b) optimisation of the excitation wavelength; (c) changes in area as a function of eluent pH value; (d) Change in resolution between muramic acid (MurN) and mannosamine (ManN) peaks and in retention time of MurN, ManN, galactosamine (GalN) and glucosamine (GlcN) as a function of tetrahydrofuran (THF) concentration in eluent (error bars represent ± standard deviation).

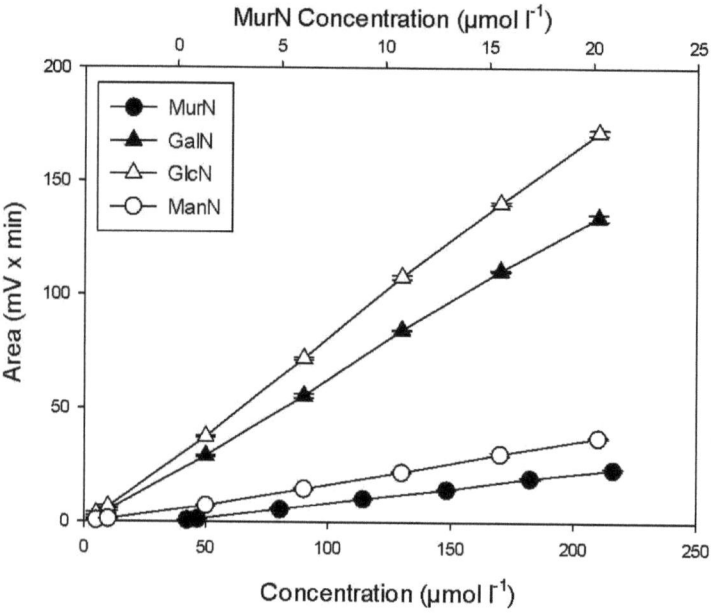

Fig. 4 Calibration curves of reference amino sugars (error bars represent ± standard deviation).

Table 3 Validation parameters of the HPLC method; intraday precision values are means of six measurements of a 90 µmol l^{-1} (muramic acid: 9 µmol l^{-1}) standard mixture analysed on the same day; inter-day precision values are means of six measurements of 90 µmol l^{-1} (muramic acid: 9 µmol l^{-1}) standard mixtures analysed on six days; recovery rates are mean values of each three test samples consisting of amino sugar standard mixture added to quartz sand samples after hydrolysis.

Component	Intraday precision	Interday precision	Correlation coefficient	Recovery rate	LOQ	LOD
	Area (mV × min) (±SD)		r	% (±SD)	mg kg^{-1}	
Muramic acid	9.13 (0.13)	9.85 (0.55)	0.999	115 (2)	4.2	1.7
Mannosamine	12.74 (0.24)	11.83 (0.75)	0.998	105 (3)	29.9	9.0
Galactosamine	55.15 (0.98)	52.83 (1.47)	0.999	112 (1)	11.9	3.6
Glucosamine	63.31 (0.68)	64.67 (3.67)	0.999	81 (1)	11.9	3.6

SD = standard deviation, LOQ = limit of quantification, defined as a signal ten times higher than blank. Limit of detection (LOD) defined as a signal three times higher than blank.

Table 4 Effect of sample solvent on the contents of muramic acid, mannosamine, galactosamine and glucosamine in soil samples

Soil	Muramic acid		Mannosamine		Galactosamine		Glucosamine	
	Water	Buffer	Water	Buffer	Water	Buffer	Water	Buffer
	$\mu g\ g^{-1}$ soil							
Forest 1	210	96	160	63	1840	800	2730	1340
Forest 2	140	110	49	16	1550	1410	2640	2300
Arable 1	140	130	33	34	690	650	1180	1170
Arable 2	117	116	<LOQ	<LOQ	840	890	1190	1360
Arable 3	110	100	<LOQ	<LOQ	910	800	1400	1280
Arable 4	67	52	<LOQ	<LOQ	540	430	980	850
Probability values								
Solvent	<0.01		<0.01		0.01		0.02	
Soil	<0.01		<0.01		<0.01		<0.01	
Soil × solvent	<0.01		<0.01		<0.01		<0.01	
CV (±%)	14	24	12	33	16	23	15	24

CV = mean coefficient of variation between replicate samples (n = 3); LOQ = limit of quantification; samples < LOQ were excluded from the ANOVA.

Table 5 Effect of sample solvent on the contents of muramic acid, mannosamine, galactosamine and glucosamine in plant litter samples.

Plant litter	Muramic acid		Mannosamine		Galactosamine		Glucosamine	
	Water	Buffer	Water	Buffer	Water	Buffer	Water	Buffer
	$\mu g\ g^{-1}$ soil							
Sugarcane filter cake	55	68	20	17	210	210	1070	880
Maize leaves	<LOQ	<LOQ	<LOQ	<LOQ	26	31	36	36
Pea leaves	<LOQ	<LOQ	34	24	16	19	98	100
Amaranth straw	42	51	23	31	170	180	1440	1550
Wheat straw	27	25	<LOQ	<LOQ	76	83	1290	1350
Solvent	0.34		0.98		0.49		0.28	
Litter	0.07		<0.01		<0.01		<0.01	
Litter × solvent	0.59		0.01		<0.01		0.86	
CV (±%)	14	27	14	18	24	18	6	10

CV = mean coefficient of variation between replicate samples (n = 3); LOQ = limit of quantification; samples < LOQ were excluded from the ANOVA.

3.4 Discussion

3.4.1 Optimisation of the mobile phase

Studies by Appuhn et al. (2004) and Zelles (1988) described a mobile phase with 0.75% THF. However, Zelles (1988) had no satisfactory separation and Appuhn et al. (2004) noted long retention times (32.4 min for glucosamine). In our study, we tested increased THF concentrations and found an optimum concentration of 1.5% (Fig. 1). This yielded in shorter retention times for all amino sugars and a slightly better resolution between muramic acid and mannosamine in comparison to Appuhn et al. (2004). Both, Hodgin (1979) and Jarret et al. (1986) had already studied the impact of THF on the amino acid separation by reversed phase chromatography. In our study, we could observe similar complex effects caused by THF. On the one hand, an increasing THF concentration caused muramic acid to change its elution position with mannosamine. On the other hand, increasing THF concentration caused better separation and shorter retention times (Fig. 3d). Roggendorf and Spatz (1981) found that THF accelerated compounds containing ether groups. This is probably the explanation for a better resolution between muramic acid and mannosamine and why muramic acid eluted faster than mannosamine with increased THF concentration, as muramic acid contains two ether groups and mannosamine only one. However, increasing THF concentration beyond 1.6%, we could not achieve reproducible peak separation anymore. For example, an occasional retention time shift for up to two minutes was observed, which caused an overlay of matrix- and amino sugar peaks. For this reason we decided to use a THF concentration of 1.5%.

Lindroth and Mopper (1979) showed that a decreasing pH value of the mobile phase yields better separation between the amino acids. However, they tested only a pH value range between 5.9 and 7.9. To examine whether a lower mobile phase pH shows a better separation, we tested a pH value range from 4 to 6 in the mobile phase. We obtained no better separation at low pH but we did observe that pH values below 5 reduced the fluorescence response (Fig. 3c) as already reported by Jarrett et al. (1986). Our results confirm the pH value chosen by Zelles (1988) and Appuhn et al. (2004).

3.4.2 Excitation wavelength optimisation

The optimal excitation wavelength is a compromise between increasing the fluorescence intensity of amino sugar peaks and decreasing fluorescence intensity of other primary amines of the matrix. For instance, at a wavelength of 280 nm a high fluorescence signal for all primary amines is observed, unfortunately also for interfering peaks like ammonia (Jarrett et al. 1986). Several studies about amino sugar determination (Mimura and Delmas 1983; Zelles 1988; Appuhn et al. 2004) use an excitation wavelength of 340 nm except for Rönköö et al. (1994), who suggested an excitation wavelength of 310 nm. But opposed to these studies, other authors like Lindroth and Mopper (1979) and Jarrett et al. (1986) recommended an excitation wavelength of 330 nm for amino acid methods. Jarrett et al. (1986) confirmed this adjustment by testing several excitation wavelengths for amino acid determination. We tested excitation wavelengths in a range between 300 nm and 340 nm to find the optimal excitation wavelength for amino sugar determination (Fig. 3a). With an optimum excitation wavelength of 330 nm, our result is consistent with that reported by Jarrett et al. (1986).

3.4.3 Ortho-phthaldialdehyde (OPA) reaction time

To obtain a high fluorescence response we tested different reaction times for the OPA-derivatisation. The fluorescence intensity of the muramic acid-isoindole-derivate was relatively constant whereas the other three amino sugars derivates decompose after 120 sec reaction time (Fig. 3b). Lindroth and Mopper (1979) suggested that electron donating groups (in our case the carboxyl-group of muramic acid) have a stabilizing effect on the isoindole group. This assumption is supported by Roth (1971), who showed in his study notably high fluorescence intensity for amino acids with two carboxyl-groups (aspartic acid, glutamic acid). Moreover, it would explain the stability and the 10 times higher fluorescence response of the muramic acid derivates. We presume that the low fluorescence response of mannosamine is closely related to the conformation of mannosamine. The primary amino-group of this molecule is located in an axial conformation, whereas the other three amino sugars contain the amino group in an equatorial conformation. Steric effects hinder the OPA reaction in mannosamine, which proceeds more easily in an equatorial conformation.

3.4.4 Validation parameters

In general, our validation parameters (Table 3) of this HPLC method are similar with validation parameters reported earlier (Zelles 1988; Appuhn et al. 2004). Zhang and Amelung (1996) described a reliable and sensitive amino sugar determination via gas chromatography. Their LOQ ranged from 10 µg ml^{-1} (muramic acid) to 20 µg ml^{-1} (other three amino sugars). In contrast to that we reached a higher sensitivity with a LOQ of 0.5 µmol l^{-1}, which is equal to 0.13 µg ml^{-1} for muramic acid, and 5.0 µmol l^{-1}, which is equal to 0.90 µg ml^{-1} for the other three amino sugars. Furthermore, with gas chromatographic methods a time-consuming off-line derivatisation procedure is needed for obtaining volatile components.

Our LOQ is similar to that of other HPLC methods described by Kaiser and Benner (2000), Diaz et al. (1996) or Ekblad and Näsholm (1996). Kaiser and Benner (2000) used a pulsed amperometric detector (PAD) whereas Diaz et al. (1996) applied a fluorescence detector with 6-aminoquinolyl-N-hydroxysuccinmydyl-carbamate (AQC) as derivatisation reagent. Ekblad and Näsholm (1996) used also fluorometric detection and 9-fluorenylmethylchloroformate (FMOC) for derivatisation. Diaz et al. (1996) determined only glucosamine and galactosamine. The peaks were well separated but each amino sugar yielded two peaks. Ekblad and Näsholm (1996) examined only glucosamine, which eluted in three peaks. Kaiser and Benner (2000) determined all four amino sugars with two different methods based on anion exchange chromatography.

3.4.5 Effect of sample solvent

Dorresteijn et al. (1996) reported that the OPA derivatisation is a very complex reaction with three possible products: hydrolysis of OPA, stabilization of OPA by 2-mercaptoethanol to yield an ortho-phthalaldehyde-2-mercaptoethanol complex and reaction of OPA with 2-mercaptoethanol and amino sugars yielding the isoindole-derivate needed for amino sugar analysis. This complexity makes it difficult to find the optimal conditions for the derivatisation. However, pH of the reaction mixture, the concentration of 2-mercaptoethanol and the reaction time are probably the most important variables to control. The effect of the reaction time is discussed above.

With regard to method accuracy, Dorresteijn et al. (1996) suggested Milli-Q water to be best suited. Appuhn et al. (2004) also used water as sample solvent in their study. However, since sample pH is between 2 and 3 in the sample extract, we examined if better accuracy for this method is possible with higher pH during OPA reaction. Sample extract pH was increased to values about 5 with phosphate-buffer. In general, buffered samples showed higher standard deviations. Significant differences between water and buffer were only observed for the soil samples, particularly for the basalt-derived clayey soil forest 1, which contained high contents of aluminium and iron oxides. The hydrolysis products of these oxides presumably form as central ions complexes in the presence of ligands like phosphate. If these complexes bind amino sugars, they may not be completely available for OPA derivatisation anymore. With respect to the higher coefficients of variation and the lower amino sugar content using buffer as sample solvent, we recommend to use water as more reliable sample solvent.

3.4.6 Concentrations of amino sugars in soil and plant material samples

Our amino sugar contents in soil (Table 4) compare well with those reported in the literature from arable and forest soils (Zelles 1988, Guggenberger et al. 1999, Joergensen et al. 2010). Guggenberger et al. (1999) obtained in different arable soils a total amino sugar composition of 64% glucosamine, 31% galactosamine, 3% muramic acid, and 2% mannosamine. Amelung et al. (1999) found in different grassland soils 60% glucosamine, 33% galactosamine, 5% muramic acid and 2% mannosamine. Liang et al. (2007a) measured in different forest soils 67% glucosamine, 21% galactosamine, 8% muramic acid, and 4% mannosamine. Ding et al. (2010) detected 55% glucosamine, 40% galactosamine, and 5% muramic acid, but no mannosamine in soil samples amended with different amounts maize residues after 38 weeks of incubation. Also Glaser et al.

(2004) did not find mannosamine in all samples. Liang et al. (2006, 2007b) measured 72% glucosamine, 8% galactosamine, 13% muramic acid and 7% mannosamine in soybean straw and maize stalks. Amelung et al. (1999), Guggenberger et al. (1999), Liang et al. (2006, 2007a/b), and Ding et al. (2010) all used after conversion of the amino sugars to aldonitrile acetates the gas chromatographic method described in detail by Zhang and Amelung (1996). Benner and Kaiser (2003) measured 63% glucosamine, 17% galactosamine, 5% mannosamine, and 16% muramic acid in marine particulate organic matter, with pulsed amperometric detection and ion exchange chromatography, a method without a derivatisation step.

It is not clear whether the absence of mannosamine in the low-biomass arable soils is due to the missing sensitivity as the majority of the data reported for mannosamine has been obtained in high biomass soils (Guggenberger et al. 1999; Turrión et al. 2002; Liang et al. 2007b). In our three soils with detectable mannosamine data, the concentration is in the range obtained by others (Amelung et al. 1999; Guggenberger et al. 1999). High mannosamine percentages, as in our pea leaves, were also one time observed in litter layer (Turrión et al. 2002).

3.5 Conclusions

Our method allows fast, quantitative, and reproducible HPLC determination of the four most important amino sugars in soil and plant litter. The separation between muramic acid and mannosamine was optimised in combination with shorter retention times in comparison with previously published methods. No interferences exist from amino acids or other primary amines, occurring in soil and plant hydrolysates.

Acknowledgments

We gratefully acknowledge Gabriele Dormann for her technical assistance. Caroline Indorf was funded by the German Research Foundation (DFG).

3.6 References

Amelung W (2001) Methods using amino sugars as markers for microbial residues in soil. In: Lal JM, Follett RF, Stewart BA (eds) Assessment methods for soil carbon. Lewis Publishers, Boca Raton, FL, pp 233-272

Amelung W (2003) Nitrogen biomarkers and their fate in soil. J Plant Nutr Soil Sci 166:677-686

Amelung W, Zhang X, Flach KW, Zech W (1999) Amino sugars in native grassland soils along a climosequence in North America. Soil Sci Soc Am J 63:6-2

Amelung W, Lobe I, Du Preez CC (2002) Fate of microbial residues in sandy soils of the South African Highveld as influenced by prolonged arable cropping. Eur J Soil Sci 53:29-35

Amelung W, Brodowski S, Sandhage-Hofmann A, Bol R (2008) Combining biomarker with stable isotope analyses for assessing the transformation and turnover of soil organic matter. Adv Agron 100:155-250

Appuhn A, Joergensen RG (2006) Microbial colonisation of roots as a function of plant species. Soil Biol Biochem 38:040-051

Appuhn A, Joergensen RG, Scheller E, Wilke B (2004) The automated determination of glucosamine, galactosamine, muramic acid and mannosamine in soil and root hydrolysates by HPLC. J Plant Nutr Soil Sci 167:17–21

Benner R, Kaiser K (2003) Abundance of amino sugars and peptidoglycan in marine particulate and dissolved organic matter. Limnol Oceanogr 48:118-128

Chantiny MH, Angers DH, Prévost D, Vézina LP, Cahlifour FP (1997) Soil aggregation and fungal and bacterial biomass under annual and perennial cropping systems. Soil Sci Soc Am J 61:262-267

Coelho RRR, Sacramento DR, Linhares LF (1997) Amino sugars in fungal melanins and soil humic acids. Eur J Soil Sci 48:425-429

Diaz J, Lliberia, JL, Comellas L, Broto-Puig F (1996) Amino acid and amino sugar determination by derivatization with 6-aminoquinolyl-N-hydroxysuccinimidyl carbamate followed by high-performance liquid chromatography and fluorescence detection. J Chromatogr A 719:171–179

Ding X, Zhang X, He H, Xie H (2010) Dynamics of soil amino sugar pools during decomposition processes of corn residues as affected by inorganic N addition. J Soils Sedim 10:758-766

Dorresteijn RC, Berwald LG, Zomer G, de Gooijer CD, Wieten G, Beuvery EC (1996) Determination of amino acids using o-phthalaldehyde-2-mercaptoethanol derivatization. Effect of reaction conditions. J Chromatogr A 724:159–167

Ekblad A, Näsholm T (1996) Determination of chitin in fungi and mycorrhizal roots by an improved HPLC analysis of glucosamine. Plant Soil 178:29-35

Engelking B, Flessa H, Joergensen RG (2007) Shifts in amino sugar and ergosterol contents after addition of sucrose and cellulose to soil. Soil Biol Biochem 39:2111-2118

Ferrero MÁ, Aparicio LR (2010) Biosynthesis and production of polysialic acids in bacteria. Appl Microbiol Biotechnol 86:1621-1635

Glaser B, Turrión MB, Alef K (2004) Amino sugars and muramic acid - biomarkers for soil microbial community structure analysis. Soil Biol Biochem 36:399-407

Guggenberger G, Frey SD, Six J, Paustian K, Elliott ET (1999) Bacterial and fungal cell wall residues in conventional and no-tillage agroecosystems. Soil Sci Soc Am J 63:1188–1198

Hodgin JC (1979) The separation of pre-column o-phthalaldehyde derivatized amino acids by high performance liquid chromatography. J Liquid Chromatog Rel Technol 2:1047-1059

Jarrett HW, Cooksy KD, Ellis B, Anderson JM (1986) The separation of o-phthalaldehyde derivates of amino acids by reversed-phase chromatography on octylsilica columns. Anal Biochem 153:189-198

Joergensen RG, Meyer B (1990) Chemical change in organic matter decomposing in and on a forest Rendzina under beech (*Fagus sylvatica* L.). J Soil Sci 41:17-27

Joergensen RG, Mäder P, Fließbach A (2010) Long-term effects of organic farming on fungal and bacterial residues in relation to microbial energy metabolism. Biol Fertil Soils 46:303-307

Kaiser K, Benner R (2000) Determination of amino sugars in environmental samples with high salt content by high-performance anion-exchange chromatography and pulsed amperometric detection. Anal Chem 72:2566-2572

Kenne LK, Lindburg B (1983) Bacterial polysaccharides. In: Aspinall GO (ed) The polysaccharides. Academic Press, New York, pp 287-353

Liang C, Zhang X, Rubert KF, Balser TC (2006) Effect of plant materials on microbial transformation of amino sugars in three soil microcosms. Biol Fertil Soils 43:631-639

Liang C, Fujinuma R, Wei LP, Balser TC (2007b) Tree species-specific effects on soil microbial residues in an upper Michigan old-growth forest system. Forestry 80:65-72

Liang C, Zhang X, Balser TC (2007a) Net microbial amino sugar accumulation process in soil as influenced by different plant material inputs. Biol Fertil Soils 44:1-7

Lindroth P, Mopper K (1979) High performance liquid chromatography determination of subpicomole amounts of amino acids by precolumn fluorescence derivatization with o-phthaldialdehyde. Anal Chem 51:1667-1674

Millar WN, Casida LE (1970) Evidence for muramic acid in soil. Can J Microbiol 16:299-304

Mimura T, Delmas D (1983) Rapid and sensitive method for muramic acid determination by high performance liquid chromatography with precolumn fluorescence derivatization. J Chromatogr 280:91-98

Probst B, Schüler C, Joergensen RG (2008) Vineyard soils under organic and conventional management – Microbial biomass and activity indices and their relation to soil chemical properties. Biol Fertil Soils 44:443-450

Roggendorf E, Spatz R (1981) Systematic use of tetrahydrofuran in reversed-phase high-performance liquid chromatography. An example of the selectivity benefits of ternary mobile phases. J Chromatogr 204:263-268

Rönkkö R, Pennanen T, Smolander A, Kitunen V, Kortemaa H, Haahtela K (1994) Quantification of Frankia strains and other root-associated bacteria in pure cultures and in the rhizosphere of axenic seedlings by high-performance liquid chromatography-based muramic acid assy. Appl Environ Microbiol 60:3672–3678

Roth M (1971) Fluorescence reaction for amino acids. Anal Chem 43:880-882

Rottmann N, Siegfried K, Buerkert A, Joergensen RG (2010) Litter decomposition in fertilizer treatments of vegetable crops under irrigated subtropical conditions. Biol Fertil Soils (DOI 10.1007/s00374-010-0501-9)

Sharon N (1965) Distribution of amino sugars in microoganisms, plants and invertebrates. In: Balasz EA, Jeanlanx RW (eds) The amino sugars, part 2A. Distribution and biological role. Academic Press, New York, pp 1-45

Stevenson F.J (1982) Organic forms of soil nitrogen. In: Stevenson FJ (ed) Nitrogen in Agricultural Soils. American Society of Agronomy, Madison, FL, pp 101-104

Turrión MB, Glaser B, Zech W (2002) Effects of deforestation on contents and distribution of amino sugars within particle-size fractions of mountain soils. Biol Fertil Soils 35:49-53

Yoneyama T, Koike, Y, Arakawa Y, Yokoyama HK, Sasaki Y, Kawamura T, Araki Y, Ito E, Takao S (1982) Distribution of mannosamine and mannosaminuronic acid among cell walls of *Bacillus* species. J Bacteriol 149:15-21

Wasylnka JA, Simmer MI, Moore MM (2001) Differences in sialic acid density in pathogenic and non-pathogenic *Aspergillus* species. Microbiol 147:869-877

Zelles L (1988) The simultaneous determination of muramic acid and glucosamine in soil by high-performance liquid chromatography with precolumn fluorescence derivatization. Biol Fertil Soils 6:125–130

Zhang X, Amelung W (1996) Gas chromatographic determination of muramic acid, glucosamine, mannosamine, and galactosamine in soils. Soil Biol Biochem 28:1201–1206

4. Comparison of HPLC methods for determination of amino sugars in soil hydrolysates

Soil Biology & Biochemistry

Caroline Indorf* [1,3], Samuel Bodé [2], Pascal Boeckx [2], Jens Dyckmans [3], Axel Meyer [4], Klaus Fischer [4], Rainer Georg Joergensen [1]

[1] Department of Soil Biology and Plant Nutrition, University of Kassel, Nordbahnhofstr. 1a, 37213 Witzenhausen, Germany

[2] Laboratory of Applied Analytical and Physical Chemistry (ISOFYS), Ghent University, Coupure Links 653, 9000 Ghent, Belgium

[3] Centre for Stable Isotope Research and Analysis, University of Göttingen, Büsgenweg 2, 37077 Göttingen, Germany

[4] Department of Analytical and Ecological Chemistry, University of Trier, Universitätsring, 54286 Trier, Germany

Abstract

A study on the suitability of chromatographic techniques such as high performance anion exchange chromatography (HPAEC) via fluorescence detection (Fl) and pulsed amperometric detection (PAD), respectively, and reversed phase (RP) chromatography for the analysis of galactosamine (GalN), glucosamine (GlcN), mannosamine (ManN) and muramic acid (MurN) in soil hydrolysates was carried out. The RP-Fl method was rapid, provided good validation parameters and relatively inexpensive instrumentation. The HPAEC methods had a slightly higher limit of quantification (LOQ) 0.6 – 5.0 µmol l^{-1} (HPAEC-Fl) and 1.0 – 10.0 µmol l^{-1} (HPAEC-PAD) in comparison to the RP-Fl method 0.5-5.0 µmol l^{-1}. Various sample pretreatment methods and some further chromatographic methods were investigated and the advantages and disadvantages of the HPLC methods are discussed.

* Corresponding author. Tel.: + 49 5542 98 1503; e-mail: cindorf@uni-kassel.de

Key words: amino sugars / HPLC / OPA / PAD / microbial residues/ HPAEC / soil analysis/ derivatisation

4. Comparison of HPLC methods for determination of amino sugars in soil hydrolysates

4.1 Introduction

The determination of amino sugars, especially MurN and GlcN, is highly suited for quantifying the fungal and bacterial contribution in soil (Joergensen and Wichern, 2008). Also Liang et al. (2007) and Amelung et al. (2008) regarded amino sugars to be the most important biomarkers for the presence of microbial residues. As plants do not contain amino sugars and soil animals contribute only minor amounts of amino sugars to soil, the majority of amino sugars are assumed to be of microbial origin (Dai et al., 2002; Appuhn and Joergensen, 2006). Amino sugars in soil are usually measured by gas chromatographic (GC) analysis (Amelung and Zhang, 2001) or reversed phase HPLC, using OPA (ortho-phthaldialdehyde) for fluorescing derivatisation (Zelles, 1988; Appuhn et al., 2004; Indorf et al., 2011). The amino sugar data published are apparently not affected by methodological differences, giving confidence in the reliability of the methods used (Joergensen and Wichern, 2008).

However, the use of compound-specific $\delta^{13}C$ analysis of amino sugars would additionally make it possible to investigate the differences in turnover between bacterial and fungal residues. Such a compound-specific $\delta^{13}C$ analysis of amino sugars is available for GC separation (Amelung and Zhang, 2001; Glaser and Amelung, 2002; Glaser, 2005; Glaser and Gross, 2005). However, the derivatisation procedure of the hydrolysis products into volatile components, e.g. alditol acetates is time consuming and needs considerable experience in laboratory work (Benzing-Purdie, 1984; Zhang and Amelung, 1996; Glaser, 2005). Moreover, the derivatisation reaction has the potential to introduce fractionation, and the introduction of external C atoms into the analysed molecule leads to a decrease in accuracy of the isotope ratio determination. An HPLC method directly linked to an isotope ratio mass spectrometer could markedly increase the sample throughput and accuracy (Bodé et al., 2009).

The RP-Fl method described by Indorf et al. (2011) is reliable, but it is not suited for adaptation with isotope ratio mass spectrometry (IRMS), due to the organic mobile phase. However, a variety of other HPLC methods are free of organic solvents, such as (1) HPAEC, (2) high performance cation exchange chromatography (HPCEC, Watanabe, 1984), and (3) high performance anion exclusion chromatography (HPEXC). Bodé et al. (2009) were the first to describe an HPAEC-method for isotope amino sugar analysis. However, this method has several drawbacks: (1) The NaOH of the mobile phase is sensitive to CO_2 contamination, (2) this alkaline mobile phase ideally requires a metal-free liquid handling system, (3) the method of Bodé et al. (2009) consisted of two separate runs, one for the basic amino sugars and one for MurN, and (4) the limit of quantification for MurN is too high for the low MurN concentrations in soil.

For these reasons, the objectives of the present study were (1) to find a reliable purification and concentration procedure for amino sugars in HCl hydrolysates, (2) to optimise the method of Bodé et al. (2009) by testing different chromatographic separation mechanisms, and (3) to test HPCEC and HPEXC as organic solvent-free alternatives for the RP-Fl method described by Indorf et al. (2011).

4.2   Materials and methods

4.2.1   Soils samples

Method evaluation was implemented using six different soil samples (0-10 cm) taken from four arable and two forest sites in Germany (Hessia and Lower Saxony), differing in physical, chemical and microbial properties (Table 1). Soil physical, chemical and biological properties of the soil samples were determined as described by Probst et al. (2008).

4.2.2   Chemicals

The reference materials (D-(+)-glucosamine (GlcN) hydrochloride, D-(+)- galactosamine (GalN) hydrochloride, D-(+)- mannosamine (ManN) hydrochloride and muramic acid (MurN) and KOH were purchased from Sigma Aldrich (St. Louis, MO, USA). The low carbonate 50 % NaOH stock solution, OPA (ortho-phthaldialdehyde), 2-mercaptoethanol, H_3BO_3, sodium acetate, sodium citrate, KH_2PO_4 and $NaNO_3$ were obtained from Merck (Darmstadt, Germany). The 6 M HCl solution, methanol and tetrahydrofurane (THF) were from VWR (West Chester PA, USA). All solutions were prepared with Milli-Q water produced via a Direct-Q 3 system (Millipore, Billerica, MA, USA). All other reagents were of high purity (≥95 %).

4. Comparison of HPLC methods for determination of amino sugars in soil hydrolysates

Table 1 Physical, chemical and microbiological properties of the soils used in this investigation

Soil	Clay	Silt	Sand	pH-H_2O	Soil organic C	Total N	Microbial biomass C	Ergosterol
	%				mg g^{-1} soil		µg g^{-1} soil	
Forest 1	15	77	8	3.9	59.9	4.3	610	3.2
Forest 2	6	39	55	3.9	58.0	2.4	520	5.4
Arable 1	35	55	10	7.0	18.6	1.8	450	1.6
Arable 2	34	56	10	7.3	15.1	1.4	360	1.1
Arable 3	18	66	16	7.8	14.0	1.2	200	0.5
Arable 4	8	8	84	7.4	7.8	0.7	180	0.6

The borate buffer solution (pH 11) was prepared by dissolving 50 g of H_3BO_3 in 900 ml of water, adjusted to pH 11 with KOH (47% solution) and diluted to 1 l with water. This solution was stable for up to 12 months at 4 °C. The reducing solution was prepared by adding 2.5 ml of 2-mercaptoethanol to 100 ml of buffer solution. This solution was stable for up to 6 months at 4 °C in the dark. The OPA-reagent for pre-column derivatisation was prepared by dissolving 25 mg of OPA in 2 ml of methanol in a brown flask, mixed with 2 ml of reducing solution and diluting to 44 ml with buffer solution. The reagent was stable for up to 7 days at 4°C in the dark. For post-column derivatisation the OPA-reagent was prepared by using the same solutions but multiplied by ten to produce 480 ml of OPA-reagent, which was covered with nitrogen during chromatography.

The standard stock solutions were prepared according to Indorf et al. (2011).

4.2.3 Amino sugar extraction

The amino sugar extraction was based on a modified method described by Appuhn et al. (2004). Air-dried soil (400 mg, sieved at 2 mm) was mixed with 10 ml of 6 M HCl. After 6-h hydrolysis at 105°C, the samples were filtered over glass filters (Whatman GF/A).

4.2.4 Purification and concentration of amino sugars

To have higher amino sugar concentration and less impurity in the soil hydrolysates, we tested the following amino sugar purification methods: (1) extraction based on Appuhn et al. (2004) as described by Indorf et al. (2011), (2) extraction based on Zhang and Amelung (1996), (3) extraction via strong cation exchange resin (AG 50W-X8, hydrogen form, 100-200 dry mesh size, Biorad, Munich, Germany), (4) solid phase extraction via strong cation exchange (SCX) cartridges (Strata SCX 500µm, 70Å, 200 mg/3 ml, Phenomenex, Aschaffenburg, Germany), (5) Extraction via strong anion exchange (SAX) cartridges (Strata SAX 500 µm, 70Å, 200 mg/3 ml, Phenomenex, Aschaffenburg, Germany) to concentrate and separate MurN from the basic amino sugars.

(1) A 0.3 ml aliquot of the filtered hydrolysate was evaporated to dryness at 40-45 °C to remove HCl, re-dissolved in water, evaporated a second time and re-dissolved in 1 ml of water. After centrifugation at 5.000 g the supernatant was frozen and stored at -18 °C until analysisA 0.3 ml aliquot was evaporated to dryness at 40-45 °C to remove HCl, re-dissolved in water, evaporated a second time and re-dissolved in 1 ml of water.

(2) For the method described by Zhang and Amelung (1996), 6 ml of the filtered hydrolysate were evaporated to dryness, re-dissolved in 20 ml of water and adjusted to a pH between 6.6 and 6.8 with

KOH(aq). After centrifugation at 5000 g for 10 min the supernatant was lyophilised. To reduce the amount of inorganic salt the residue was re-dissolved in 3 ml of dry methanol and centrifuged at 5000 g for 10 min. The clear supernatant was dried under a stream of air, re-dissolved in 5 ml of water, filtered over a 45 µm filter and stored in vials at − 18 °C until analysis.

(3) The sample preparation via cation exchange resin was performed by using a strong cation exchange resin. 2 ml of the protonated resin suspended in water were filled into plastic tubes (Biorad), the resin was activated with 2x 5 ml of 0.1M HCl and washed with 5 ml of water. 5 ml of the filtered sample hydrolysate was evaporated to dryness and re-dissolved in 1 ml of water. Both steps were repeated once and the sample was subsequently transferred into 2 ml tubes (Eppendorf) and centrifuged at 5.000 g. Then the resin was loaded with the complete supernatant. To obtain a quantitative transfer of the complete sample, the pellet was re-dissolved in 1 ml of water, centrifuged again and the supernatant was added onto the resin. For elution of the neutral and negatively charged compounds, the resin was washed with 3 x 4 ml of water. Subsequently the analytes were eluted with 3 x 4 ml of 0.5 M HCl. The eluate was evaporated to dryness and re-dissolved in 1 ml of water. After centrifugation at 5.000 g, the supernatant was transferred into vials and stored at − 18 °C.

(4) For the SCX extraction the SPE cartridges were conditioned with 2.4 ml of methanol and equilibrated with 2.4 ml of a 0.1 M KH_2PO_4 buffer solution at pH 6. After loading the sample hydrolysate (5 ml, adjusted to a pH between 2 and 4 with $KOH_{(aq)}$) onto the SPE cartridge, the cartridge was washed with an 8 % aqueous methanol solution. The analytes were eluted with 3 x 2 ml of 5 % methanolic NH_3. The eluate was evaporated to dryness at 40 °C, re-dissolved in water, evaporated a second time and re-dissolved in 1 ml of water. After centrifugation at 5.000 g, the supernatant was frozen and stored at -18 °C until analysis.

(5) For the SAX-extraction, the SPE cartridge was conditioned with 2.4 ml of methanol and equilibrated with 2.4 ml of 300 mM NaOH followed by 2.4 ml of 2 mM NaOH. Then 5 ml of the sample (adjusted to a pH between 5 and 7 with $KOH_{(aq.)}$) were loaded onto the cartridge. After washing the cartridge with 3 x 2.4 ml of 2 mM NaOH the analytes were eluted with 3 x 2.4 ml of a 4 mM $NaNO_3$ + 4 mM NaOH (50/50 v/v) solution. The eluate was lyophilised, re-dissolved in 1 ml of water and kept at -18 °C.

4.2.5 Reversed phase-HPLC and pre-column derivatisation

The amino sugar determination via reversed phase and pre-column derivatisation was carried out as described by Indorf et al. (2011). Chromatographic separations were performed on a Phenomenex (Aschaffenburg, Germany) Hyperclone C_{18} (ODS) column (125 mm length × 4 mm diameter, 5 µm particle size, 12 nm pore size), protected by a Phenomenex C_{18} guard cartridge (4 mm length × 2 mm diameter). The HPLC system consisted of a Dionex (Germering, Germany) P 580 gradient pump, a Dionex Ultimate WPS – 3000TSL analytical autosampler with in-line split-loop injection and thermostat and a Dionex RF 2000 fluorescence detector set at 445 nm emission and 330 nm excitation wavelengths with medium sensitivity. Further chromatographic conditions are shown in Table 2.

For the automated pre-column derivatisation, 50 µl of OPA and 30 µl of sample were mixed in the preparation vial and after 120 sec reaction time 15 µl of the indole derivates were injected. The mobile phase consisted of two eluents. Eluent A was a 97.8/0.7/1.5 (v/v/v) mixture of a buffer solution, methanol and tetrahydrofuran (THF). The buffer solution contained 52 mmol sodium citrate and 4 mmol sodium acetate, adjusted to pH 5.3 with HCl. Then methanol and THF were added. Eluent B consisted of 50% water and 50% methanol (v/v). The amino sugar separation was performed isocratically, but a gradient was used for cleaning the column after each run. Every run was started at an eluent A/B v/v composition of 93/7 for 19 min. A linear gradient was run to reach 80% B after 3 min and remaining isocratic for 3 min. A reverse gradient to 93/7 within 3 min was followed by 2 min isocratic run after which the column was preconditioned for the next sample.

4.2.6 HPAEC and post column derivatisation

The chromatographic conditions were based on those of Bodé et al. (2009) and consisted of two different methods for the basic amino sugars and for the acid amino sugar MurN, respectively. The anion exchange separations for both methods were performed using a Dionex CarboPac PA20 column (150 mm length x 3 mm diameter, 6.5 µm particle size) protected by a Dionex AminoTrap guard column (30 mm length x 3 mm diameter). The HPAEC system consisted of a metal-free PEEK gradient pump (Sykam S2100, Fürstenfeldbruck, Germany), a PEEK autosampler (Sykam S5200) and a Dionex RF 2000 fluorescence detector set at 445 nm emission and 330 nm excitation wavelength with medium sensitivity. For the post-column derivatisation a knitted reactor was placed between the column outlet and the detector inlet. The OPA derivatisation reagent was delivered by a Sykam S1122 micro pump. By means of a Y-piece, which was placed between column outlet and knitted reactor, the OPA reagent was continuously mixed with the column eluent.

In the knitted reactor the separated amino sugars reacted with the OPA to form the indole derivatives that were detected by fluorescence.

The mobile phase for GalN, ManN and GlcN consisted of 2 mM NaOH. Since carbonate negatively affects the amino sugar separation, the NaOH eluent was degassed by sonication under vacuum for 15 min and was covered with nitrogen during chromatography. After every run the column was cleaned with 200 mM NaOH for 5 minutes and equilibrated for the next sample with 2 mM NaOH for 15 minutes. In addition, after every six runs, the column had to be cleaned with 200 mM NaOH for 30 min and then equilibrated with 2 mM NaOH for a further 30 minutes.

To obtain reproducible results for MurN it was necessary to use the on-line purification described by Bodé et al. (2009). Therefore the injected sample was first passed to a purification column (Dionex CarboGuard, 30 mm length x 3 mm diameter). There, the sample was washed with 2 mM NaOH for 3 min. Then MurN was eluted from the purification column with an eluent consisting of 25 % 100 mM sodium acetate + 100 mM NaOH / 70 % H_2O / 5 % 200 mM NaOH onto the analytical column (CarboPac PA 20) protected by a Dionex AminoTrap guard column (30 mm length x 3 mm diameter) for chromatographic separation. MurN eluted after approx. 60 min and subsequently the purification column was cleaned with 200 mM NaOH for 2 min and equilibrated with 2 mM NaOH for 3 min before the next sample was injected. It was necessary to clean the column with 200 mM NaOH for 30 min und equilibrate with 25 % 100 mM sodium acetate + 100 mM NaOH / 70 % H_2O / 5 % 200 mM NaOH for 30 min after every six runs. Further chromatographic conditions are listed in Table 2.

4. Comparison of HPLC methods for determination of amino sugars in soil hydrolysates

Table 2 Chromatographic conditions of the different methods

Methods	HPAEC-PAD	HPAEC-Fl (basic amino sugars)	HPAEC-Fl (MurN)	HPCEC-Fl	RP-Fl
Separation column	CarboPac PA20	CarboPac PA20	CarboPac PA20	Luna SCX	Hyperclone C_{18} (ODS) column
Guard column	AminoTrap + Carbopac PA 20 guard column	AminoTrap	AminoTrap + Carbopac PA 20 guard column as purification column	SecurityGuard	C_{18} guard cartridge
Eluent composition	Gradient (NaOH/ sodium acetate/ water)	Isocratic (2 mM NaOH)	Isocratic (NaOH/ sodium acetate/ water)	Isocratic (10 mM KH_2PO_4, pH 7)	Gradient (buffer/methanol/THF)
Column temperature (°C)	20	15	30	50	35
Flow rate (ml min^{-1})	0.30	0.35	0.40	0.50	1.5
Injection volumne (µl)	15	15	15	10	15 (indole derivate)
Derivatisation	No derivatisation	Post column derivatisation (OPA-flow : 0.10 ml min^{-1})	Post column derivatisation (OPA-flow : 0.16 ml min^{-1})	Post column derivatisation (OPA-flow : 0.10 ml min^{-1})	Pre- column derivatisation

4. Comparison of HPLC methods for determination of amino sugars in soil hydrolysates

Table 2 Chromatographic conditions of the different methods

Methods		HPAEC-PAD	HPAEC-Fl (basic amino sugars)	HPAEC-Fl (MurN)	HPCEC-Fl	RP-Fl
t_R (min)	GalN	20.1	15.8	-	14.8	15.4
	ManN	22.9	17.9	-	16.8	9.4
	GlcN	26.3	20.6	-	17.2	16.6
	MurN	42.3	-	60.1	8.2	8.4

4.2.7 HPAEC and pulsed amperometric detection (PAD)

Amino sugar analysis based on Meyer et al. (2008) was performed with a Dionex ion chromatography system consisting of an autosampler AS 50, a gradient pump GP 40 and an electrochemical detector ED 40 equipped with a thin-layer-type amperometric cell. The detector cell integrated a gold working electrode and an Ag / AgCl reference electrode. Chromatographic separation was achieved using a CarboPac PA20 column (150 mm length × 3 mm diameter, 6.5 µm particle size) coupled with a CarboPac PA 20 guard column (30 mm length × 3 mm diameter) and an AminoTrap guard column (30 mm length × 3 mm diameter).
Sample and standard vials were held at 10 °C in the autosampler.

The separation of GalN, ManN, GlcN and MurN was carried out using 1 M sodium acetate + 25mM NaOH (eluent A), 250 mM NaOH (eluent B), 10 mM NaOH (eluent C) and water (eluent D). Table 3 describes the gradient used for separation of amino sugars. The analytes were detected by applying a quadrupole-potential waveform on the gold electrode. The potentials were E1 = 0.1 V from 0 to 0.4 ms, E2 = 2.0V from 0.41 to 0.42 ms, E3 = 0.6V from 0.42 to 0.43 ms and E4 = - 0.1V from 0.44 to 0.50 ms. Table 2 shows further chromatographic conditions.

4.2.8 HPCEC and post-column derivatisation

The cation exchange separations, based on Watanabe [16], were conducted on a Phenomenex (Aschaffenburg, Germany) LUNA SCX column (150 mm length × 4.66 mm diameter, 5 µm particle size) which was preceded by a Phenomenex SCX security guard cartridge (4 mm length × 2 mm diameter). The HPCEC system and the post column derivatisation methodology were identical to those described above for the HPAEC method with post column derivatisation. 10 µl of the sample aliquots were reacted with OPA after chromatographic separation and quantified by fluorescence as described above. Further chromatographic conditions are shown in Table 2.

Table 3 Gradient profile for amino sugar separation by HPAEC-PAD

Time (min)	% A (1 M sodium acetate / 25 mM NaOH)	% B (250 mM NaOH)	% C (10 mM NaOH)	% D (water)
0	0	0	20	80
27	0	0	20	80
28	5	20	0	75
29	5	20	0	75
42	30	20	0	50
44	60	20	0	20
46	60	20	0	20
48	0	50	0	50
50	0	50	0	50
52	0	100	0	0
60	0	100	0	0
68	0	0	20	80
90	0	20	0	80

4.2.9 HPEXC and post column derivatisation

For obtaining an alternative method of the MurN determination described by Bodé et al. (2009) we also tested the HPEXC. This method has the advantage of running with an acidicbuffereluent. The HPEXC analyses were performed on the chromatographic fluorescence system as described for the HPAEC method. The separation of MurN was accomplished on a Phenomenex Rezex Roa-Organic acid (300 mm length × 7.8 diameter, 8 µm particle size) using 1% H_3PO_4 as mobile phase and a column temperature of 25°C. The flow rate was set to 0.6 ml min^{-1} and the injection volume was 15 µl. The post-column derivatisation methodology was according to the one described above. After chromatographic separation and OPA reaction the indole derivates were detected by fluorescence as described above.

4.3 Results

4.3.1 Sample purification methods

We investigated the purification and the concentration of soil and tested different sample preparation methods. For four purification methods tested the interfering matrix substances could not be substantially eliminated from the amino sugars. However, amino sugar concentrations could be increased without increasing the matrix concentrations by using strong cation exchange purification. The strong cation exchange purification via Biorad *AG50W X8* resin (method 3) and the SPE-SCX purification (method 4) gave the highest amounts for all four amino sugars, without an increase in impurities (Fig. 1). Both purification methods are based on the strong cation exchange mechanisms. The resins of these methods have the same functional groups (sulfonic acid) but the matrix of the resins is different. At method 3 the sulfonic group is attached to a styrene divinyl benzene copolymere matrix, whereas at method 4 the benzene sulfonic acid group is bonded to the surface of silica particle. Further, method 3 and 4 differ in eluent solutions used. For evaluating the purification of the different sample purification methods we determined the area of the matrix in a defined section relative to the area of the analyte peaks. The matrix area reached 34 % (method 1) and 32 % (method 2) of that of the analytes, respectively. A very low matrix area (2 % and 4 %, respectively, related to the signal sizes of the basic amino sugars) results from the use of the extraction methods 3 and 4. The matrix percentage of the MurN chromatogram is 93% and 98% for method 1 and 2, respectively, and 92% for method 3 and 4. No amino sugars were recovered via method 5.

4. Comparison of HPLC methods for determination of amino sugars in soil hydrolysates

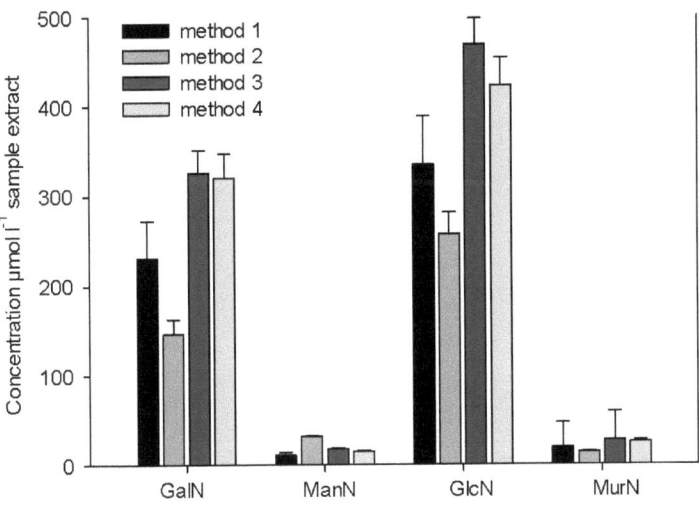

Fig. 1 Comparison of four different sample purification methods (1) based on Appuhn et al. (2004), (2) based on Zhang and Amelung (1996), (3) sample purification via a strong cation exchange resin and (4) solid phase extraction via strong cation exchange cartridges, respectively of GalN, ManN, GlcN and MurN in a soil extract (forest 1) (error bars represent ± standard deviation). *Note*: Amino sugar concentrations presented are soil extract concentrations and not sample concentrations. The concentrations were measured via HPAEC-Fl.

4.3.2 Chromatographic methods

4.3.2.1 HPAEC analysis with fluorescence detection

As described by Bodé et al. (2009), the HPAEC separation mechanism made it possible to determine GalN, ManN, GlcN and MurN in standard solutions as well as in soil extracts. Fig. 2c shows the resulting chromatogram with baseline resolution between each of the basic compounds, whereas Fig. 2d shows the optimum result for the MurN method. Table 2 shows the retention times (t_R) by HPAEC fluorescence detection for the basic amino sugar method as well as for the acidic amino sugar method.

4. Comparison of HPLC methods for determination of amino sugars in soil hydrolysates

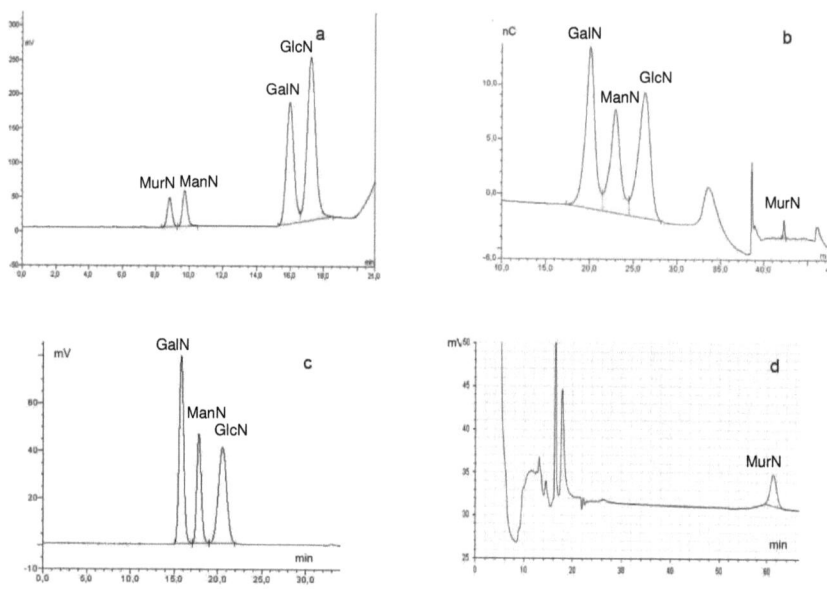

Fig. 2 Chromatograms of a standard mixture consisting of 130 µmol l^{-1} GalN, 130 µmol l^{-1} ManN, 130 µmol l^{-1} GlcN and 13 µmol l^{-1} MurN (a) obtained from the RP-Fl method and (b) obtained from the HPAEC-PAD method; chromatograms of (c) a standard mixture consisting of 130 µmol l^{-1} GalN, 130 µmol l-1 ManN, 130 µmol l-1 GlcN measured by HPAEC-Fl for basic amino sugars and of (d) a standard solution containing 13 µmol l^{-1} muramic acid obtained from the acidic HPAEC-Fl method.

4. Comparison of HPLC methods for determination of amino sugars in soil hydrolysates

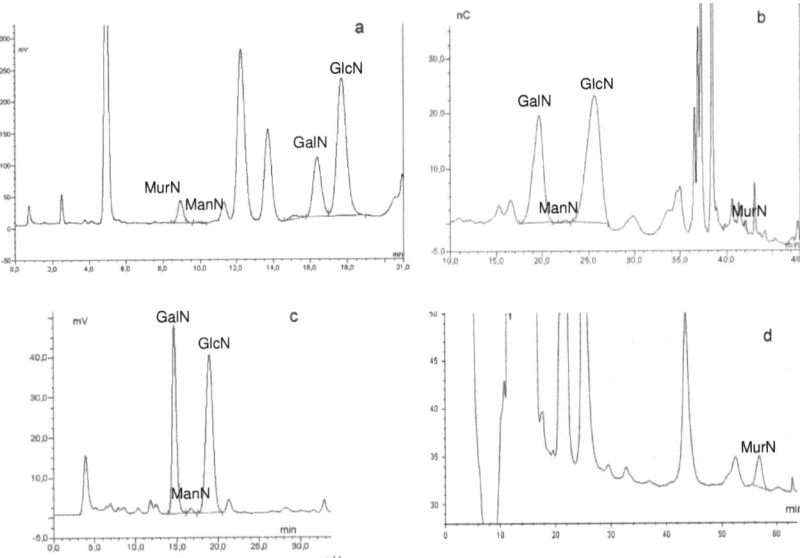

Fig. 3 Chromatograms of a soil hydrolysate (forest 1) (a) measured by RP-Fl, (b) obtained from the HPAEC-PAD method (c) obtained from the HPAEC-Fl method for basic amino sugars and (d) obtained from the acidic HPAEC-Fl method.

The highest fluorescence response for the amino sugar standards was yielded at an OPA flow rate of 0.10 ml min^{-1} for the basic amino sugars and 0.16 ml min^{-1} for MurN, respectively. The coefficient of variation (CV) was roughly between 3% and 6% for intraday and between 3% and 6% for interday precision, respectively (Table 4a). . The calibration curves were linear in a range from 5 to 210 µmol l^{-1} for the basic amino sugars and 1 to 21 µmol l^{-1} for MurN, with good correlation coefficients and standard deviations (Table 4b, Fig. 4). This reflects the reproducibility and the precision of the analytical method excluding sample preparation. The lowest quantification levels (LOQ, 10 times higher than noise) of the amino sugar standards were between 0.6 µmol l^{-1} for MurN and 5 µmol l^{-1} for ManN (Table 4b).

Table 4a Validation parameters of the RP-Fl, HPAEC-Fl and HPAEC-PAD methods; intraday precision values are mean coefficient of variation (CV) between replicate samples (n = 3) of three measurements of a 90 µmol l^{-1} (MurN: 9 µmol l^{-1}) standard mixture analysed on the same day; interday precision values are mean CV of three measurements of 90 µmol l^{-1} (MurN: 9 µmol l^{-1}) standard mixtures analysed on three days

Component	Intraday Precision (Coefficient of variation)			Interday Precision (Coefficient of variation)		
	RP-Fl	HPAEC-Fl	HPAEC-PAD	RP-Fl	HPAEC-Fl	HPAEC-PAD
GalN	1.8%	2.8%	1.0%	2.8%	2.6%	44.5%
ManN	1.9%	6.3%	1.7%	4.2%	3.4%	40.9%
GlcN	1.1%	5.7%	1.6%	3.1%	2.8%	48.4%
MurN	0.1%	5.6%	6.1%	9.2%	5.9%	48.4%

4. Comparison of HPLC methods for determination of amino sugars in soil hydrolysates

Table 4b+c Validation parameters of the RP-Fl, HPAEC-Fl and HPAEC-PAD methods

Component	LOQ (μmol l^{-1} sample extract)			LOD (μmol l^{-1} sample extract)		
	RP-Fl	HPAEC-Fl	HPAEC-PAD	RP-Fl	HPAEC-Fl	HPAEC-PAD
GalN	2.0	3.0	5.0	0.6	0.9	1.5
ManN	5.0	5.0	10.0	1.5	1.5	3.0
GlcN	2.0	3.0	6.0	0.6	0.9	1.8
MurN	0.5	0.6	1.0	0.2	0.3	0.3

Component	R^2 (Correlation Coefficient)			Slope of calibration line		
	RP-Fl	HPAEC-Fl	RP-Fl	HPAEC-Fl	RP-Fl	HPAEC-Fl
GalN	0.999	0.999	0.999	0.999	0.999	0.999
ManN	0.998	0.999	0.998	0.999	0.998	0.999
GlcN	0.999	0.999	0.999	0.999	0.999	0.999
MurN	0.999	0.997	0.999	0.997	0.999	0.997

LOQ = limit of quantification, defined as a signal ten times higher than noise. Limit of detection (LOD) defined as a signal three times higher than noise. SD = standard deviation

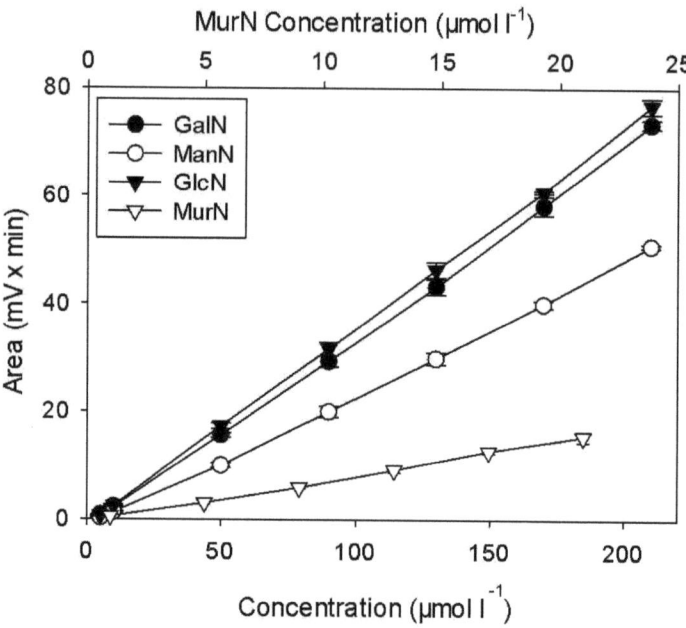

Fig. 4 Calibration curves of reference amino sugars obtained from HPAEC-Fl (error bars represent ± standard deviation).

4.3.2.2 HPAEC analysis and pulsed amperometric detection

The HPAEC-PAD method system provided reproducible separation and quantification of GalN, ManN, GlcN and MurN (Fig. 2b). The elution order and retention times by HPAEC-PAD are shown in Table 2. The intraday CV for the amino sugars was between 2% and 6%. The interday CV was between 41% and 48% (Table 4a). Fig. 5 shows the curves and demonstrates the linearity of response for each of the analytes ($R^2 \geq 0.992$) (Table 4b). The calibration curves were linear between 10 µmol l^{-1} and 210 µmol l^{-1} for the basic amino sugars and between 1 µmol l^{-1} and 21 µmol l^{-1} respectively (Fig. 5). The LOQ depended on the amino sugar and varied from 1µmol l^{-1} for MurN to 10 µmol l^{-1} for ManN (Table 4b).

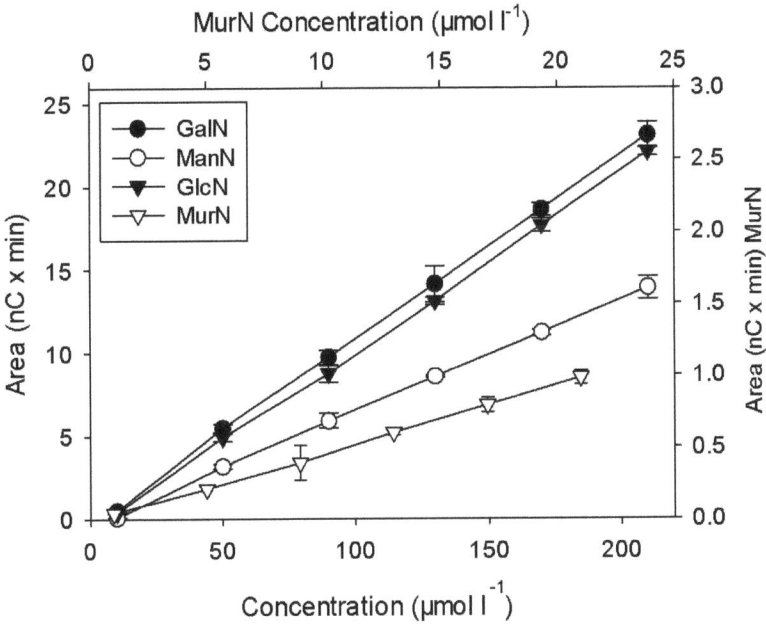

Fig. 5 Calibration curves of reference amino sugars obtained from HPAEC-PAD (error bars represent ± standard deviation).

4.3.2.3 HPCEC and HPEXC analysis and fluorescence detection

We did not obtain satisfactory separation with the HPCEC mechanism for the four amino sugars. Because of the acidic group, which is fully deprotonated at pH 7, MurN was very poorly retained on the cation exchange column. However, it was well separated from the other amino sugars in the standard solution but it coeluted with matrix peaks in sample extracts.

As expected, the HPEXC mechanism was only applicable for MurN. The basic amino sugars did not elute from the column because of their high affinity to the HPEXC column material. The determination of MurN (peak area and retention time) was not reproducible, due to the low affinity of MurN to the stationary phase.

4.3.2.4 Amino sugar contents of soil with regard to the different HPLC methods

The content of the different amino sugars in soils increased in the order ManN < MurN < GalN < GlcN (Table 5), regardless of the HPLC method used. In three soils the ManN content was below the LOQ for all methods tested (Table 5). On average, the highest amino sugar values were measured using HPAEC-Fl, except for ManN. The GlcN content contributed between 57% (HPAEC-Fl) and 62% (HPAEC-PAD) to the total amino sugar content. The contribution of GalN and MurN to the total amino sugar content varied between 34% (HPAEC-PAD) and 36% (HPLC-RP), and 3% (HPAEC-PAD) and 5% (HPAEC-Fl), respectively, whereas ManN, if present in measurable concentration, contributed between 1% (HPAEC-Fl) and 2% (HPLC-RP) to the total amino sugar content. This shows that all chromatographic methods resulted in a similar pattern of amino sugars in soil samples.

The different HPLC methods had no significant effect on the concentration of ManN and GlcN measured in soil samples. However, the concentrations measured by the HPAEC-PAD were significantly smaller than those measured by the RP-Fl and HPAEC-Fl methods (Table 5).

4. Comparison of HPLC methods for determination of amino sugars in soil hydrolysates

Table 5a Comparison of amino sugar results in soil samples obtained by RP-HPLC, HPAEC-Fl and HPAEC-PAD

Soil	ManN			GalN		
	RP-HPLC	HPAEC-FL	HPAEC-PAD	RP-HPLC	HPAEC-FL	HPAEC-PAD
	μg g^{-1} soil					
Forest 1	160	120	160	1840	2340	1920
Forest 2	49	23	14	1550	1570	1080
Arable 1	34	31	14	690	890	860
Arable 2	<LOQ	<LOQ	<LOQ	840	1020	680
Arable 3	<LOQ	<LOQ	<LOQ	910	920	440
Arable 4	<LOQ	<LOQ	<LOQ	540	350	310
Probability values						
Method	0.30			<0.01		
Soil	<0.01			<0.01		
Soil × Method	0.37			0.01		
CV (±%)	12	30	25	16	18	9

4. Comparison of HPLC methods for determination of amino sugars in soil hydrolysates

Table 5b Comparison of amino sugar results in soil samples obtained by RP-HPLC, HPAEC-Fl and HPAEC-PAD

Soil	GlcN			MurN		
	RP-HPLC	HPAEC-FL	RP-HPLC	HPAEC-FL	RP-HPLC	HPAEC-FL
			$\mu g\ g^{-1}$ soil			
Forest 1	2730	3120	2730	3120	2730	3120
Forest 2	2640	2750	2640	2750	2640	2750
Arable 1	1180	1680	1180	1680	1180	1680
Arable 2	1190	1400	1190	1400	1190	1400
Arable 3	1400	1280	1400	1280	1400	1280
Arable 4	980	620	980	620	980	620
Probability values						
Method	0.20		0.20		0.20	
Soil	<0.01		<0.01		<0.01	
Soil × Method	0.07		0.07		0.07	
CV (±%)	15	21	15	21	15	21

4.4 Discussion

4.4.1 Sample purification methods

The amino sugar extraction method described by Appuhn et al. (2004) (method 1) provided good results for the reversed-phase method (Indorf et al., 2011). However, our objective was to concentrate MurN and not the matrix substances, in order to determine MurN using detectors with a lower sensitivity for MurN, such as an isotope ratio mass spectrometer after HPLC. First we tried to concentrate the sample extract obtained from method 1 by taking 1 ml sample hydrolysate instead of 0.3 ml. But this led to a higher matrix concentration, which aggravated the separation and degraded the column material. Owing to these problems, we tested different extraction methods (Fig. 1) and found the strong cation exchange extraction (method 3) to be best suited for methods with a high LOQ for MurN. Method 3 is most suitable for reducing the matrix in sample extracts. Previous studies (Amelung and Zhang, 2000; Kaiser and Benner, 2000) have indicated that cation exchange resins are useful for eliminating interference compounds and obtaining higher amounts of amino acids and amino sugars, respectively. For the extraction we used a strong cation exchange resin, which was composed of sulfonic acid functional groups attached to a styrene divinylbenzene copolymer lattice. The negatively charged sulfonic acid groups of the resin bound the positively charged amino groups of the amino sugars until a stronger eluent removed the amino sugars. Interfering compounds that are neutral or negatively charged were eliminated beforehand by washing the resin with water. With this method we obtained a higher concentration of amino sugars without an increase of the matrix substances in the analytical aliquot, particularly for the basic amino sugars (Fig. 1). The concentration of MurN in the analytical aliquot could only be increased marginally. This is caused by the carboxyl group of muramic acid, which leads to a low affinity of MurN to the resin material.

4.4.2 HPAEC-Fl

A baseline separation for the three basic amino sugars was achieved using the HPLC parameters suggested by Bodé et al. (2009). As described earlier (Bodé et al., 2009), other concentrations of the NaOH eluent, other column temperatures and other flow rates decreased the selectivity of the amino sugar chromatography. Previous studies (Bodé et al., 2009; Boschker et al., 2008; Meyer et al., 2008; Yu et al., 2003) have indicated that carboxylic acids like MurN do not elute with NaOH as the mobile phase because its carboxyl group has a very high affinity to the column material. Thus an eluent with a higher ionic strength is needed for the elution of MurN.

Several authors (Cataldi et al., 1999; Lee, 1996; Lu et al., 1997; Rocklin et al., 1983; Wong and Jane, 1995) have suggested sodium acetate and $NaNO_3$, respectively, as an applicable "pressing agent", whereas sulfate and carbonate are not suitable (Lee, 1996; Rocklin et al., 1983).In addition, we tested phosphate as a pushing anion, which even at low concentrations led to a co-elution of all compounds, due to its high ionic strength. Cataldi et al. (1999) have claimed that nitrate as an additional competing ion for the elution of sugar acids leads to shorter retention times at millimolar concentrations, but acetate salts perform better, due to their lower ionic strength. We could not use nitrate as an eluent because Calhoun et al. (1983) and Chen (1973) showed that nitrate acts as a quencher affecting the FI detection. Consequently, we chose acetate as a pressing agent.

The use of a gradient for obtaining MurN together with the basic amino sugars in one run was not possible, due to the fact that MurN was overlayed by peaks of the matrix. Only the online purification described by Bodé et al. (2009) provided good and reproducible results for MurN. However, similarly to the method for the basic amino sugars, a loss of activity of the column was observed after 4 h, which was shown by insufficient separation. Therefore, regeneration of the column with a mobile phase consisting of 25% 100 mM sodium acetate and 75% 200 mM NaOH for 15 min and a subsequent re-equilibration for 40 min was necessary after every 3 runs. As already mentioned by Bodé et al. (2009), the decreasing activity of the column is due to the accumulation of stronger binding ions like carbonates on the column. The post column derivatisation performed well.

4.4.3 HPAEC-PAD

The advantages of the HPAEC-PAD methodology are the lack of need for derivatisation, the simultaneous determination of all 4 analytes in one run and the robustness against matrix interferences. We used this method for testing whether the OPA derivatisation had an effect on the measured amino sugar concentrations. For a comparison of the methods, it was necessary to use a similar mobile phase and, consequently, we used the method based on Meyer et al. (2008). This method had to be modified to obtain a baseline separation of the three basic amino sugars at the beginning of the run. Therefore, we reduced the column temperature from 35°C to 20°C. The pressure and the retention times increased by lowering the column temperature, but the separation of the three basic amino sugars significantly improved.

4.4.4 HPCEC and HPEXC

Watanabe (1984) described a determination of MurN via cation exchange chromatography and amperometric detection. He used KH_2PO_4 at pH 2.7 as the eluent. At this low pH-value we obtained two peaks for each amino sugar analyte. We obtained single peaks from pH 5 up to 7. Through an increase of the pH to 7 and of the column temperature to 50 °C a better but not optimal resolution of the basic amino sugars was achieved. A decrease of the KH_2PO_3 concentration from 30 mM to 20mM increased the retention times (Watanabe, 1984) but did not improve the resolution of the amino sugars, whereas an increase of the KH_2PO_3 concentration from 30 mM to 40 mM led to poor separation. This method is not suitable for amino sugar determination in soil hydrolysates, because the MurN peak could not be separated from matrix substances.

HPEXC was tested as an additional alternative for the determination of MurN. The Rezex ROA anion exclusion column was only applicable with water or mineral acids as the eluent. We tested water, H_3PO_4, and H_2SO_4 at different column temperatures (20-86 °C) and different concentrations (0-250 mM H_2SO_4). MurN eluted only with 1% phosphoric acid as the mobile phase, whereas the basic amino sugars did not even elute after five hours. The separation of the ion exclusion mechanism is mainly based on the Donnan membrane exclusion principle (Hicks, 1985; Fischer, 2002). Hick et al. (1985) described ion exclusion as a process where ionic solutes pass through the interstitial volume of the resin, but are prevented from entering the pore volume of the sulfonated divenylbenzene column material, because of the Donnan membrane effects. At an anion exclusion stationary phase, only sample anions are excluded from the resin phase by the fixed charges of the sulfonate groups of the cation exchange resin (Fritz, 1991). In mineral acids, basic amino sugars are protonated and positively charged, which would explain the high affinity to the column material. Probably a stronger cationic eluent, such as K_2HPO_4, would elute the basic amino sugars. However, the column material is only suitable for mineral acids and water. The amino group of MurN is also protonated, whereas the carboxyl group is probably partially negatively charged, which causes a very low affinity of MurN to the column material. So we assume that MurN was probably not retained at all, because the retention time is almost identical to the dead time of the HPLC system. In conclusion, the anion exclusion separation mechanism is not suitable either for MurN or for basic amino sugar determination. For determining amino sugars via HPEXC a derivatisation, e.g. an acetylation of the amino group, would probably be necessary.

4.4.5 Comparison of RP-Fl, HPAEC-Fl and HPAEC-PAD

On RP-Fl, it was possible to determine all four amino sugars simultaneously within 26 min (Indorf et al., 2011), whereas HPAEC-PAD analysis required 90 min. A simultaneous determination of all four amino sugars via HPAEC-Fl was not achievable. The reversed phase separation with pre-column derivatisation is more suitable for a simultaneous determination of the basic amino sugars and MurN, because after derivatisation all four amino sugars are neutral indole derivatives with similar chemical properties, which makes it possible to use an isocratic method. In HPAEC without pre-column derivatisation the deprotonated carboxyl group of MurN has a much stronger affinity towards the positively charged column material than the basic amino sugars. This makes the use of a gradient or the implementation of two separate runs indispensable. Furthermore, because of the many matrix substances contained in soil, a gradient with a very slight slope has to be used. Other studies using anion exchange chromatography show similar results. Cataldi et al. (1999) and Meyer et al. (2008) determined amino sugars, sugars and uronic acids simultaneously by using a gradient, whereas Benner and Kaiser (2000) as well as Bodé et al. (2009) determined MurN in a separate chromatographic run.

For HPAEC-Fl a gradient could not achieve a separation between MurN and the matrix peaks. This is probably due to the significantly broader matrix peaks in comparison to the small matrix peaks detected by HPAEC-PAD. In HPAEC-Fl a separate run for MurN is necessary, because the MurN peak was otherwise completely overlapped by matrix peaks.

RP-Fl of amino sugar analysis was rapid and relatively inexpensive compared with the HPAEC method, which required expensive columns, permanent helium or nitrogen stripping of the eluent, a time-consuming column cleaning and ideally a metal-free PEEK-consisting HPLC system. Although HPAEC is more time consuming and expensive than RP-Fl, the validation parameters of these methods were basically the same for the three methods. The sensitivity was good for the RP-Fl method, with a LOQ between 0.5 µmol l^{-1} sample extract (MurN) and 5.0 µmol l^{-1} sample extract (ManN), the LOQ of the HPAEC-Fl was slightly higher, whereas the LOQ of HPAEC-PAD was even higher than the HPAEC-Fl LOQ (Table 3b). This shows that the sensitivity of fluorescence detection in combination with OPA derivatisation is similar to that of PAD.

The better correlation coefficients and the lower standard deviations of the precision values (Table 4a) of the fluorescence methods in comparison with the PAD-method are probably caused by the good baseline separation. A full baseline separation was not obtained for the HPAEC-PAD method. For achieving a baseline separation, we tried lowering the column temperature to 15°C. As a consequence, the retention times increased, but without obtaining an improvement in separation. Theoretically, a longer equilibration (e.g. 60 min) at the end of each run would improve separation.

There were no significant differences between the HPAEC-Fl and the RP-Fl method with regard to the amounts of all four amino sugars in soil hydrolysates (Table 5). As described above, we obtained partly significantly different amino sugar amounts in soil hydrolysates by using HPAEC-PAD. The reason for that is the different detection, not the chromatography. Probably the remaining matrix has an influence on the quantification by adsorbing on the PAD electrode surface. This would explain the lower sensitivity as well as the very low interday precision (Table 4a). As a result, both separation mechanisms (RP and HPAEC) provided similar results for amino sugars in soil samples. Thus, we have confirmed that the HPAEC separation mechanism gives valid results for amino sugar specific $\delta^{13}C$-analysis in soil samples, as shown by Bodé et al. (2009).

4.5 Conclusions

On the basis of the data from this study, we conclude that the reversed phase method described by Indorf et al. (2011) is a reliable, rapid, inexpensive method best suited for amino sugar determination in plant and soil hydrolysates. When using HPAEC as the chromatographic system, a pulsed electrochemical detection is recommended. In comparison to fluorescence detection and post-column derivatisation, the HPAEC-PAD method has 3 advantages. (1) No derivatisation is necessary, (2) the matrix interferences are lower and (3) no separate run for determining MurN is required. However, a disadvantage of the HPAEC-PAD method is the somewhat high LOQ in comparison to the HPAEC-Fl method (Table 4b). The sample purification method 1 based on Appuhn et al. (2004) and modified by Indorf et al. (2011) is rapid and provide good results for sensitive chromatographic systems. For samples with very low amino sugar content or a less sensitive chromatographic system, such as LC-IRMS, sample purification via strong cation exchange resin is recommended. However, the HPAEC method as described by Bodé et al. (2009) is suitable for the determination of amino sugars in soil hydrolysates and can be used for ^{13}C-specific amino sugar detection. Further research is required on increasing concentration and purification of MurN in soil samples.

Acknowledgments

We gratefully acknowledge Gabriele Dormann, Reinhard Langel and Marion Wacht for their technical assistance. Caroline Indorf was funded by the German Research Foundation (DFG).

4.6 References

Amelung, W., Zhang, X., 2000. Determination of amino acid enantiomers in soils. Soil Biol. Biochem. 33, 553-562.

Amelung, W., Zhang, X., 2001. Determination of amino acid enantiomers in soil. Soil Biol. Biochem. 33, 223-562.

Amelung, W., Brodowski, S., Sandhage-Hofmann, A., Bol, R., 2008. Combining biomarker with stable isotope analyses for assessing the transformation and turnover of soil organic matter. Adv. Agron. 100, 155-250.

Appuhn, A., Joergensen, R.G., Scheller, E., Wilke, B., 2004. The automated determination of glucosamine, galactosamine, muramic acid and mannosamine in soil and root hydrolysates by HPLC. J. Plant Nutr. Soil Sci. 167, 17–21.

Appuhn, A., Joergensen, R.G., 2006. Microbial colonisation of roots as a function of plant species. Soil Biol. Biochem. 38, 040-051.

Benzing-Purdie, L., 1984. Amino sugar distribution in four soils as determined by high resolution gas chromatography. Soil Sci. Soc. Am. J. 48, 219-222.

Bodé, S., Denef, K., Boeckx, P., 2009. Development and evaluation of a high-performance liquid chromatography isotope ratio mass spectrometry methodology for $\delta^{13}C$ analysis of amino sugars in soil. Rapid Comm. Mass Spec.. 23, 2519-2526.

Boschker, H.T.S., Moerdijk-Poortvliet, T.C.W., van Breugel, P., Houtekamer, M., Middelburg, J.J., 2008. A versatile method for stable carbon isotope analysis of carbohydrates by high-performance liquid chromatography/ isotope ratio mass spectrometry. Rapid Comm. Mass Spec. 22, 3902-3908.

Calhoun, D.B., Vanderkooi, J.M., Englander, S.W., 1983. Penetration of small molecules into proteins studied by quenching of phosphorescence and fluorescence. Biochem. 22, 1533-1539.

Cataldi, T.R.I., Campa, C., Casella, I.G., 1999. Study of sugar acids separation by high-performance anion-exchange chromatography-pulsed amperometric detection using alkaline eluents spiked with Ba^{2+}, Sr^{2+}, Ca^{2+} as acetate or nitrate salts. J. Chromatogr. A 848, 71-81.

Chen, R.F., 1973. Quenching of the fluorescence of proteins by silver nitrate. Arch. Biochem. Biophys. 158, 605-622.

Dai, X.Y., Ping, C.L., Hines, M.E., Zhang, X.D., Zech, W., 2002. Amino sugars in artic soils. Comm. Soil Sci. Plant Anal. 33, 789-805.

Fischer, K., 2002. Environmental analysis of aliphatic carboxylic acids by ion-exclusion chromatography. Anal. Chim. Acta 465, 157–173.

Fritz, J.S., 1991. Principles and applications of ion-exclusion chromatography. J. Chromatogr. 546, 111–118.

Glaser, B., Amelung, W., 2002. Determination of $\delta^{13}C$ natural abundance of amino acid enantiomers in soil: methodological considerations and first results. Rapid Comm. Mass Spec. 16, 891-898.

Glaser, B., 2005. Compound-specific stable-isotope ($\delta^{13}C$) analysis in soil science. J. Plant Nutr. Soil Sci. 168, 633-648.

Glaser, B., Gross, S., 2005. Compound-specific $\delta^{13}C$ analysis of individual amino sugars - a tool to quantify timing and amount of soil microbial residue stabilization. Rapid Comm. Mass Spec. 19, 1409-1416.

Hicks, K.B., Lim, P.C., Haas, M.J., 1985. Analysis of uronic and aldonic acids, their lactones, and related compounds by high-performance liquid chromatography on cation-exchange resins. J. Chromatogr. 319, 159-171

Indorf, C., Dyckmans, J., Khan, K.S., Joergensen, R.G., 2011. Optimisation of amino sugar quantification by HPLC in soil and plant hydrolysates. Biol. Fertil. Soils (DOI 10.1007/s00374-011-0545-5).

Joergensen, R.G., Wichern, F., 2008. Quantitative assessment of the fungal contribution to microbial tissue in soil. Soil Biol. Biochem. 40, 2977-2991.

Kaiser, K., Benner, R., 2000. Determination of amino sugars in environmental samples with high salt content by high-performance anion-exchange chromatography and pulsed amperometric detection. Anal. Chem. 72, 2566-2572.

Lee, Y.C 1996. Carbohydrate analyses with high-performance anion-exchange chromatography. J. Chromatogr. A 720, 137-149.

Liang, C., Fujinuma, R., Wie, L.P., Balser, T.C., 2007. Tree species-specific effects on soil microbial residues in an upper Michigan old-growth forest system. Forestry 80, 65-72.

Lu, T., Lacourse, W.R., Jane, J., 1997. Evaluating sodium salts as pushing agents on high-performance anion-exchange chromatography with pulsed amperometric detection for maltodextrin analysis. Starch 49, 505-511.

Meyer, A., Fischer, H., Kuzyakov, Y., Fischer, K., 2008. Improved RP-HPLC and anion-exchange chromatography methods for the determination of amino acids and carbohydrates in soil solutions. J. Plant Nutr. Soil Sci. 171, 917-926.

Probst, B., Schüler, C., Joergensen, R.G., 2008. Vineyard soils under organic and conventional management – Microbial biomass and activity indices and their relation to soil chemical properties. Biol. Fertil. Soils 44, 443-450.

Rocklin, R.D., Pohl, C.A., 1983. Determination of carbohydrates by anion exchange chromatography with pulsed amperometric detection. J. Liq. Chrom. Relat. Tech. 6(9), 1577-1590.

Yu, H., Ding, Y.S., Mou, S.F., 2003. Direct and simultaneous determination of amino acids and sugars in rice wine by high-performance anion-exchange chromatography with integrated pulsed amperometric detection. Chromatographia 57, 721-728.

Watanabe, N., 1984. Amperometric detection of muramic acid in high-performance liquid chromatography with a post-column reaction. J. Chromatogr. 316, 495-500.

Wong, K.S., Jane, J., 1995. Effects of pushing agents on the separation and detection of debranched amylopectin by high-performance anion-exchange chromatography with pulsed amperometric detection. J. Liq. Chrom. Relat. Tech. 18, 63-80.

Zelles L (1988) The simultaneous determination of muramic acid and glucosamine in soil by high-performance liquid chromatography with precolumn fluorescence derivatization. Biol Fertil Soils 6:125–130

Zhang, X., Amelung, W., 1996. Gas chromatographic determination of muramic acid, glucosamine, mannosamine, and galactosamine in soils. Soil Biol. Biochem. 28, 1201–1206.

5. Determination of saprotrophic fungi turnover in different substrates by glucosamine-specific $\delta^{13}C$ liquid chromatography/isotope ratio mass spectrometry

Fungal ecology

Caroline Indorf*[1,2], Felix Stamm[1], Jens Dyckmans[2], Rainer Georg Joergensen[1]

[1] Department of Soil Biology and Plant Nutrition, University of Kassel, Nordbahnhofstr. 1a, 37213 Witzenhausen, Germany

[2] Centre for Stable Isotope Research and Analysis, University of Göttingen, Büsgenweg 2, 37077 Göttingen, Germany

ABSTRACT

A high performance anion exchange chromatography (HPAEC) isotopic ratio mass spectrometry (IRMS) method was developed for amino sugar-specific $\delta^{13}C$ analysis in plant hydrolysates. The HPAEC-IRMS method provided good validation parameters and the amino sugar concentrations were similar to those obtained by reversed phase (RP) high performance liquid chromatography (HPLC) and fluorescence (Fl) detection. The limit of quantification (LOQ) was 150 µmol l^{-1}. This optimised HPAEC-IRMS method opens up the possibility of a glucosamine (GlcN) specific $\delta^{13}C$ analysis in plant material. Thus, it was possible to determine the $\delta^{13}C$ values in newly formed fungal GlcN for the first time. The formation and turnover of saprotrophic fungi was investigated by using the improved HPAEC-IRMS method for GlcN-specific $\delta^{13}C$ analysis. The cultivation of saprotrophic fungi (*Lentinula edodes* and *Pleurotus* species) showed the preferred formation of fungal biomass from maize-derived (80%) rather than from beech wood-derived C. The results indicate a faster formation of fungal biomass from maize than from wheat straw as co-substrate.

Key words: amino sugars / HPAEC / HPLC-IRMS / isotope / saprotrophic fungi

* Corresponding author. Tel.: + 49 5542 98 1503; e-mail: cindorf@uni-kassel.de

5. Determination of saprotrophic fungi turnover by glucosamine-specific $\delta^{13}C$ LC-IRMS

5.1 Introduction

The contribution of saprotrophic fungi to the biogeochemistry of terrestrial ecosystems is enormous (Högberg et al, 1999). Owing to their decomposer properties, these fungi have a high impact in biotechnological degrading processes. The cultivation of saprotrophic, wood-decomposing fungi on organic waste is an efficient technique for recycling waste products like agricultural residues (Stamets 2000). Further, edible saprotrophic fungi play an important role in food industries (Leatham 1982, Stamets 2000) and traditional medicine (Lee & Balick 2006).

Little is known about the turnover of saprotrophic fungi. For a faster growth and a more efficient use of the profitable decomposer properties of the saprotrophic fungi, a better understanding of the formation and turnover of these fungi is indispensable. The natural abundance of stable isotopes expressed as δ-value is particularly suitable for performing tracer experiments and therefore for determining the formation and turnover of microorganisms. Based on the fact that C_3 and C_4 plants have different $\delta^{13}C$ values due to their different photosynthetic mechanisms, the use of C_4 and C_3 substrates as suitable tracers is presented in various studies (Engelking *et al.* 2007ab, Engelking *et al.* 2008, Semenina & Tiunov 2010) Advanced analytical methods like compound-specific $\delta^{13}C$ analysis by HPLC-IRMS simplify the determination of fungal turnover. The GC-C-IRMS method often described in the literature for amino sugar-specific $\delta^{13}C$ analysis is time consuming and causes isotope fractionation, due to derivatisation. Decock *et al.* (2009) suggested HPAEC-IRMS as a promising alternative, since derivatisation is not necessary for this method. Glucosamine (GlcN) is an indicator for fungal biomass (Ekblad & Näsholm 1995; Appuhn & Joergensen 2006) and therefore GlcN-specific $\delta^{13}C$ analysis facilitates a determination of the origin of newly formed fungal biomass.

The objectives of the present paper were (1) to adapt the HPAEC-IRMS method for amino sugar determination in soil hydrolysates based on Bodé *et al.* (2009) to an amino sugar analysis in plant and fungi samples and (2) to determine the formation and turnover of saprotrophic fungi, particularly the origin of the newly formed fungal biomass by HPAEC-IRMS.

5. Determination of saprotrophic fungi turnover by glucosamine-specific δ^{13}C LC-IRMS

5.2 Materials and methods

5.2.1 Material

The optimisation of the HPAEC-IRMS method was implemented using three different plant litter samples (Table 1). The fungal growth experiment was carried out using three different saprotrophic wood-decomposing fungi (*Lentinula edodes* P., *Pleurotus ostreatus* K. and *Pleurotus citrinopileatus* S.). All fungi were obtained as grain spawn from Mycelia bvba (Nevele, Belgium). A mixture of beech (*Fagus sylvatica* L.) wood sawdust + maize (*Zea mays* L.) straw and a mixture of beech wood sawdust and wheat (*Triticum aestivum* L.) straw, respectively, were used as fungal substrates. The chemical properties of fungi and substrates are shown in Table 2.

Table 1 HPAEC-IRMS method optimisation: Comparison of galactosamine (GalN) and glucosamine (GlcN) results in plant samples obtained by RP-HPLC and HPAEC-IRMS. Main effects of the one-way ANOVA using plant material as independent factor and method as repeated measures for amino sugar amounts.

Plant material	GalN		GlcN	
	RP-HPLC	HPAEC-IRMS	RP-HPLC	HPAEC-IRMS
	(μg g^{-1} dry matter)			
Amaranth straw	170	170	1440	1490
Wheat straw	76	82	1290	1210
Maize leaves	26	15	36	66
Probability values				
Method	NS		NS	
Plant material	<0.01		<0.01	
Plant material × method	NS		NS	
CV (±%)	24	27	6	6

Table 2 Chemical properties of the material used in the fungus experiment

Material	MurN	ManN	GalN	GlcN	δ^{13}C GlcN	δ^{13}C bulk	Total N	Total C
	µg g^{-1} dry matter				‰		%	
Wheat	39	26	99	1100	-26.6	-27.5	0.8	43.0
Maize	4	7	3	245	-12.5	-12.7	1.3	43.4
Beech wood	20	16	13	250	-26.1	-26.0	0.4	47.0
Lentinula edodes.	-	82	3	19400	-27.0	-26.0	4.4	37.3
Pleurotus ostreatus.	-	103	9	22900	-26.6	-26.9	4.5	41.5
P. citrinopileatus	-	175	2	16600	-26.8	-26.8	4.5	41.2
CV	13	7	40	5	1	1	12	2

CV = pooled coefficient of variation

5.2.2 Solutions

All solutions were prepared with Milli-Q water produced via a Direct-Q 3 system (Millipore, Billerica, MA, USA). All other reagents were of high purity (≥95 %). The derivatisation reagents and standard stock solutions were prepared according to Indorf *et al.* (2011). The low carbonate 50% NaOH solution was purchased from Merck.

5.2.3 Substrate preparation and spawning

The experiment was carried out with 6 treatments (2 different substrates × 3 different fungal strains) in 4 replicates, sampled at five time points. Consequently, 120 cultivation substrates were prepared for this experiment, 60 substrates consisted of beech wood and maize straw and the other 60 substrates were made of beech wood and wheat straw. The substrate components were mixed in special experimental jars (Microbox 0118/80+OD 118, Combiness, Eke-Nazareth, Belgium), each containing 15 g beech wood sawdust and 15 g maize straw or 15 g beech wood sawdust and 15 g wheat straw. All jars were supplied with 30 g water, that of the wood-wheat mixture additionally contained 41.06 mg $(NH_4)_2SO_4$ to obtain the same C/N ratio as the wood-maize-mixture. After mixing, the substrate-containing jars were autoclaved at 121 °C for 20 min.

The two different substrates were both inoculated with 3 g of grain spawn using 3 different fungal species (*Lentinula edodes* P., *Pleurotus ostreatus* K. and *Pleurotus citrinopileatus* S.). Then, the samples were weighed and incubated at 24 °C for 4 weeks in a 0.48 m^3 incubator, provided with 0.24 m^3 h^{-1} water filtered fresh air. The samples in the incubator were randomly distributed, each replicate was assigned to a separate grid. Twenty-four were taken initially and after every week.

5.2.4 Amino sugar extraction

The amino sugar extraction was based on the method described by Appuhn *et al.* (2004). Oven-dried (60°C for 4-5 days) plant material (800 mg, ball-milled) was mixed with 10 ml of 6 M HCl. After 3 h hydrolysis at 105°C, the samples were filtered over glass filters (Whatman GF/A). A 0.3 ml aliquot was evaporated to dryness at 40-45 °C to remove HCl, re-dissolved in water, evaporated a second time and re-dissolved in 1 ml water. After centrifugation at 5000 g, the supernatant was frozen and stored at -18 °C until analysis.

5.2.5 Amino sugar determination

The amino sugar determination by reversed phase and pre-column derivatisation was based on the method described by Indorf *et al.* (2011). Chromatographic separations were performed on a Phenomenex (Aschaffenburg, Germany) Hyperclone C$_{18}$ (ODS) column (125 mm length × 4 mm diameter, 5 µm particle size, 12 nm pore size), protected by a Phenomenex C$_{18}$ guard cartridge (4 mm length × 2 mm diameter). The column was kept at 35°C. The HPLC system consisted of a Dionex (Germering, Germany) P 580 gradient pump, a Dionex Ultimate WPS–3000TSL analytical autosampler with in-line split-loop injection and thermostat and a Dionex RF 2000 fluorescence detector set at 445 nm emission and 330 nm excitation wavelengths with medium sensitivity. For the automated pre-column derivatisation, 50 µl OPA and 30 µl sample were mixed in the preparation vial and after 120 sec reaction time 15 µl of the indole derivatives were injected. The mobile phase consisted of two eluents. Eluent A was a 97.8/0.7/1.5 (v/v/v) mixture of a water phase, methanol and tetrahydrofuran (THF). The water phase contained 52 mmol sodium citrate and 4 mmol sodium acetate, adjusted to pH 5.3 with HCl. Then methanol and THF were added. Eluent B consisted of 50% water and 50% methanol (v/v).

5.2.6 Amino sugar specific $\delta\ ^{13}C$ analysis by HPAEC-IRMS

The chromatographic conditions described by Bodé *et al.* (2009) had to be modified for the analysis of plant material. The anion exchange separation was performed using a Dionex CarboPac PA20 column (150 mm length x 3 mm diameter, 6.5 µm particle size) protected by a Dionex AminoTrap guard column (30 mm length x 3 mm diameter). The HPLC system consisted of a metal-free PEEK gradient pump (Sykam S2100, Fuerstenfeldbruck), a metal-free Peek autosampler (Sykam S5200), a LC Isolink interface (Thermo Electron, Bremen) and an isotope ratio mass spectrometer (IRMS, DELTA^{PLUS} XP, Thermo Electron).

The mobile phase consisted of 1 mM NaOH and the flow rate was set to 0.30 ml min^{-1}. The NaOH eluent was degassed by sonication under vacuum for 15 min and was covered with N_2 during chromatography. GlcN and galactosamine (GalN)) analysis was performed isocratically. The columns were maintained at 38 °C. The injection volume was 20 µl. After chromatographic separation, the eluent was delivered to the LC Isolink interface, where all organic components were quantitatively oxidised to CO_2 (Krummen *et al.* 2004). This oxidation was performed by mixing $Na_2S_2O_8$ (50 µl min^{-1}) and H_3PO_4, (50 µl min^{-1}) into the eluent, directed to an oxidation reactor set at 99.9 °C. Then, a helium stream stripped the CO_2 produced from the eluent into a capillary gas exchanger. The uncovered gas was directed over a gas dryer to the IRMS. After every run the column was cleaned with 200 mM NaOH for 15 min and equilibrated for the next sample with 1 mM NaOH for 30 min.

5.2.7 Total C and total $\delta\ ^{13}C$

Total C and total $\delta\ ^{13}C$ in the oven-dried, ball milled bulk samples were measured on a Delta plus IRMS 251 (Finnigan Mat, Bremen, Germany) after combustion using a Carlo Erba NA 1500 elemental analyser.

5.2.8 Calculations and statistical analysis

The fungal C was calculated by multiplying fungal GlcN by 9 (Appuhn & Joergensen 2006). The new fungal C was obtained by subtracting the fungal C from the fungal C contained in the sample of week 0.

Fungal C_{new} = (fungal $GlcN_0 \times 9$) - (fungal $GlcN \times 9$) [1]

The relative specific allocation (RSA) is the proportion of newly incorporated maize derived GlcN (f_{new}) relative to total glucosamine in the sample (Cliquet et al. 1990; Dyckmans et al. 2000). The RSA and the maize-derived fungal GlcN were calculated as follows:

$$f_{Maize} = \frac{(\delta GlcN_{SubstrateM} - \delta GlcN_{SubstrateW})}{(\delta GlcN_{Maize} - \delta GlcN_{Wheat})} \quad [2]$$

$$RSA(\%) = \frac{(f_{Maize} \times total\ GlcN\ (g)) - (f_{Maize} f MaizeGlcN_{T_0} \times total\ GlcN_{T_0}(g))}{Total\ GlcN(g) - (total\ GlcN_{T_0}(g) - fungal\ GlcN_{T_0}(g))} \quad [3]$$

where δ-GlcN$_{maize}$ was the GlcN-specific δ^{13}C value of pure maize straw, δ-GlcN$_{wheat}$ was the GlcN-specific δ^{13}C value of pure wheat straw, whereas δ-GlcN$_{substrateM}$ was the GlcN-specific δ^{13}C value of the maize-wood-fungi mixture and δ-GlcN$_{SubstrateW}$ was the GlcN-specific δ^{13}C value of the wheat-wood-fungi mixture. Total GlcN is the total GlcN concentration of the maize-wood-fungi mixture, total GlcN$_{T0}$ is the total GlcN concentration of the maize-wood-fungi sample at week 0. The fungal GlcN is the part of GlcN in the sample that was provided by the fungi at week 0.

The significance of differences between substrates and fungi was tested by a week-specific two-way ANOVA or by two-way ANOVA using sampling week as repeated measures. Statistical analyses were carried out using SPSS statistical software (SPSS 19.0). The results presented in tables and figures are arithmetic means and are given on an oven-dry basis.

5.3 Results

5.3.1 HPAEC-IRMS method optimisation

The GlcN peak of a standard mixture, using the HPAEC-IRMS method described by Bodé et al. (2009), eluted at 1680 sec (Fig. 1a), which moved to 720 sec after changing the eluent concentration and the column temperature (Fig. 1b). The GlcN peak of a wheat straw hydrolysate, using the HPAEC-IRMS method described by Bodé et al. (2009), coeluted with a matrix peak of the plant material (Fig. 1c), whereas the GlcN peak was clearly separated from the matrix by our optimised method (Fig. 1d). The δ^{13}C values of GalN and GlcN in a standard solution were reproducible from a concentration of 200 μmol l^{-1}. In a concentration range between 200 μmol l^{-1} and 900 μmol l^{-1} the coefficients of variation (CV) were between 0.2 and 1.4 for the δ^{13}C values of GalN and GlcN in a standard solution mix, whereas the lowest CV were obtained at concentrations between 300 and 700 μmol l^{-1}. The calibration lines for GalN and GlcN resulted in correlation coefficients (r^2) between 0.998 and 0.999, in a linear concentration range from 200 μmol l^{-1} to 900 μmol l^{-1}. The lowest quantification levels (LOQ, 10-times higher than noise) of the GalN and GlcN

5. Determination of saprotrophic fungi turnover by glucosamine-specific $\delta^{13}C$ LC-IRMS

standard solution were 50 µmol l^{-1}. However, to obtain reliable $\delta^{13}C$ values, a concentration of 200 µmol l^{-1} is necessary.

A comparison of this optimised HPAEC-IRMS method with the validated reversed phase HPLC method provided similar results for the concentrations of GalN and GlcN in different plant materials (Table 1). The CV for the GalN- and GlcN-specific $\delta^{13}C$ values ranged between 1 for GalN and 2 for GlcN.

5.3.2 The fungal growth experiment

The concentration of total C generally decreased during the incubation, without any significant effect of the fungal species, but with a significantly stronger decline in the maize-wood mixture (Table. 3). The GlcN concentration increased significantly from 1070 µg g^{-1} dry matter (wheat-wood mixtures) and 620 µg g^{-1} dry matter (maize-wood mixtures), respectively, at week 0 to 2200 µg g^{-1} dry matter (wheat-wood mixtures) and 4000 µg g^{-1} dry matter (maize-wood mixtures) at week 4, respectively. Nevertheless, the maize-wood mixtures, especially those inoculated with *Lentinula edodes,* showed a higher increase in GlcN than the wheat-wood mixtures (Table 4). The maize-wood mixtures revealed a continuous increase of newly formed fungal C over the weeks (Fig. 2).

The average bulk $\delta^{13}C$ values over the whole incubation period were -27 ‰ for the wheat-wood and -21 ‰ for the maize-wood mixtures. Significant differences were observed between the substrates but not between the weeks or the different fungi by a two-way ANOVA using the week as repeated measures (results not shown). The situation was different for the GlcN-specific $\delta^{13}C$ values of the maize-wood mixtures, which increased significantly on average from -17.5 to -14.2‰ (Table 4). This means that they were considerably higher than the bulk $\delta^{13}C$ values. After 1 week, 60% of the newly formed fungal GlcN was maize derived, and roughly 80% after 2 weeks. No further increase in maize-derived GlcN was observed after 3 and 4 weeks (Fig. 3).

5. Determination of saprotrophic fungi turnover by glucosamine-specific $\delta^{13}C$ LC-IRMS

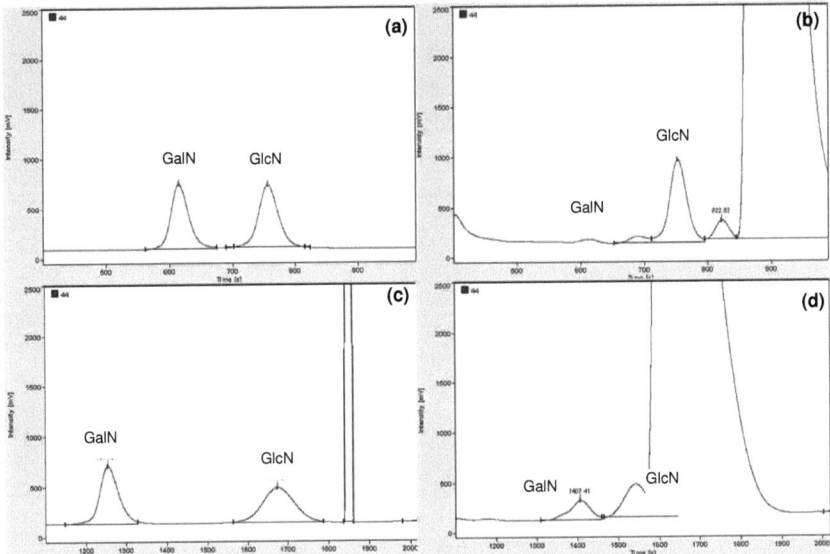

Fig. 1 Chromatograms of (a) a standard mixture consisting of 200 µmol l^{-1} GalN and 200 µmol l^{-1} GlcN and (b) a wheat straw hydrolysate, both obtained from the optimised HPAEC-IRMS method and chromatograms of (c) standard mixture consisting of 200 µmol l^{-1} GalN and 200 µmol l^{-1} GlcN and (d) a wheat straw hydrolysate, both measured by HPAEC-IRMS method described by Bodé et al. (2009).

5. Determination of saprotrophic fungi turnover by glucosamine-specific δ^{13}C LC-IRMS

Table 3 Total C in wheat-wood and maize-wood mixtures inoculated with *Lentinula edodes*, *Pleurutus ostreatus*, and *Pleurotus citrinopileatus*, respectively, during 4 weeks incubation; main effects of the 2-way week specific ANOVA, using substrate and fungi as independent factors and week as repeated measures.

Week	Total C (%)
0	45.66
1	45.22
2	45.04
3	44.98
4	44.75
Probability values	
Substrate	<0.01
Fungi	NS
Week	0.01
Substrate × fungi	0.01
Substrate × week	0.03
Fungi × week	NS
CV (±%)	1

5. Determination of saprotrophic fungi turnover by glucosamine-specific δ^{13}C LC-IRMS

Table 4 Glucosamine (GlcN) amounts and GlcN-specific δ13C values in wheat (W) and maize (M) samples inoculated with Lentinula edodes (LE), Pleurutus ostreatus (PO) and Pleurotus citrinopileatus singer (PC), respectively, during 4 week incubation; main effects of the 2-way week specific ANOVA using substrate and fungi as independent factors.

	GlcN (µg g^{-1} dry matter)					δ^{13}C GlcN (‰)				
Week	0	1	2	3	4	0	1	2	3	4
Sample										
LE-W	1020	1100	1370	2090	2520	-26.5	-27.4	-26.7	-26.3	-26.3
PO-W	1110	1220	1800	2000	2070	-26.7	-26.6	-26.3	-26.2	-26.2
PC-W	1090	1300	1680	1870	1970	-26.3	-26.4	-26.0	-25.9	-25.8
LE-M	660	970	2000	3490	5030	-18.6	-17.6	-15.6	-14.7	-14.1
PO-M	660	1430	2920	3050	3370	-18.9	-17.8	-15.4	-14.7	-14.5
PC-M	550	1240	2780	3250	3610	-17.7	-16.6	-14.5	-14.1	-13.9
Substrate	<0.01	NS.	<0.01	<0.01	<0.01	<0.01	<0.01	<0.01	<0.01	<0.01
Fungi	NS		<0.01	<0.01	NS	<0.01	<0.01	<0.01	0.02	<0.01
Substrate × fungi	NS	NS	NS	0.03	NS	NS	0.03	NS	0.04	NS
					<0.01					
CV (±%)	10	17	17	10	17	3	3	3	2	2

CV = pooled coefficient of variation; NS = not significant.

5. Determination of saprotrophic fungi turnover by glucosamine-specific δ^{13}C LC-IRMS

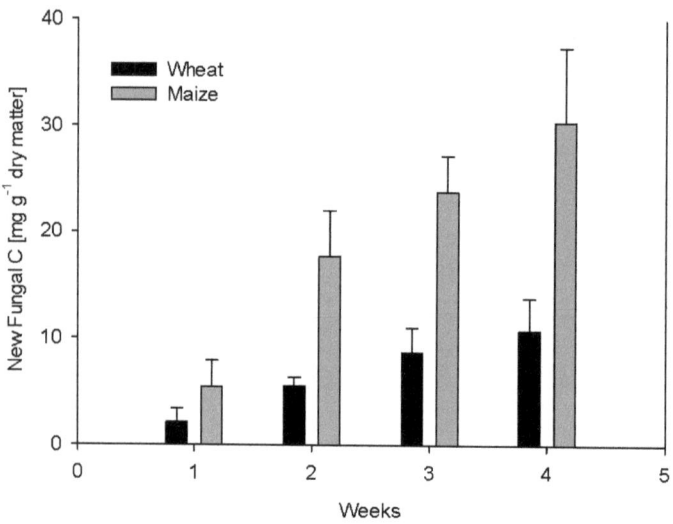

Fig. 2 Formation of new fungal C (calculated according equation (1); one bar represents the mean value of all three fungi strains) in wheat-wood and maize-wood samples inoculated with fungi during 4 week incubation at 24°C; error bars represent the standard deviation, n=12.

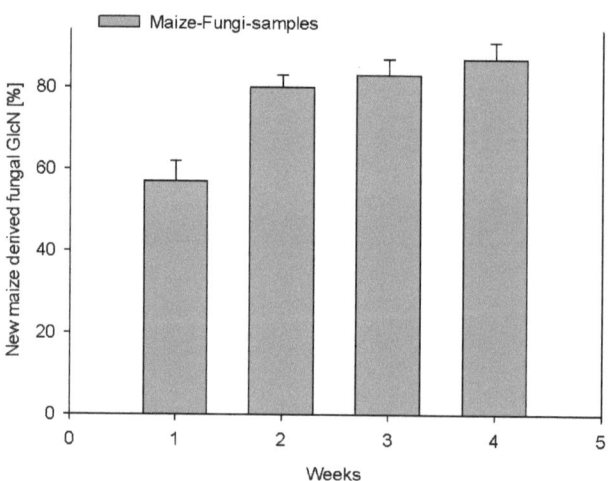

Fig. 3 Formation of new maize derived GlcN (calculated according equation (2) and (3); one bar represents the mean value of all three fungi strains) in maize-wood samples inoculated with fungi during 4 week incubation at 24°C; error bars represent the standard deviation, n=12.

5. Determination of saprotrophic fungi turnover by glucosamine-specific δ^{13}C LC-IRMS

5.4 Discussion

5.4.1 HPAEC-IRMS optimisation

The optimised amino-sugar specific δ^{13}C HPAEC-IRMS method is suitable for the determination of GalN and GlcN in soil and plant hydrolysates. This is an advantage in comparison with the method described by Bodé et al. (2009), which performed well for soil samples, but not for plant material, as the matrix of the plant samples strongly interfered with GlcN. This advantage is of special importance as the amino sugar dynamics are certainly stronger in freshly colonised plant samples than in the bulk soil (Appuhn et al. 2006).

A determination of amino sugars by HPLC-Fl and post-column derivatisation with ortho-phthaldialdehyde (OPA) of the different plant samples also revealed the presence of this enormous matrix peak for all tested plant materials. Since OPA only reacts with primary amines, this interfering matrix must be a compound that, similarly to the amino sugars, contains an amine as a functional group. Amino acids can be excluded as interfering compounds, due to the use of an amino trap. All amino acids remained on this trap and were only eluted if a stronger eluent like nitrate was used and, even in this case, not before 60 min on the column. The interfering plant matrix components may be amines that originate from transamination during the decomposition of plant materials (Mannheim et al. 1996). The interfering components might be products of the Maillard reaction, a complex series of non-enzymatic reactions leading to hundreds of end products (melanoidins) and many stable intermediates like glycosylamines (Pigman et al. 1959, Nursten 2005). The Maillard reaction occurs between an amine (e.g. protein, peptide or amino acid) and a carbonyl compound (e.g. reducing sugar). The formation of some Maillard products is acid catalysed (Nursten 2005), which implies that the acid hydrolysis of the samples may have accelerated some steps of the Maillard reaction.

A purification and elimination of the matrix peaks with cation exchange resins resulted in no improvement. Only the increase of the column temperature from 15°C to 38°C yielded a clearly better separation. Horvath et al. (1967) already described the effect of temperature on the separation by using ion exchange chromatography. They showed in their studies that the column performance improved with temperature, due to the increased solute diffusivity in the stationary as well as in the mobile phase. They also observed a significant reduction in tailing. The increase in temperature causes a reduction of the capacity factor, which is a criterion for how much longer the sample molecules stay in the stationary phase than in the mobile phase.

A low capacity factor reduces retention time but decreases selectivity (Camman 2001). Consequently, the analyte determines which temperature gives the best results. In our case, a higher temperature improved the separation of GlcN from the plant matrix components, but impaired the separation between GalN and GlcN. At 38°C column temperature, a good compromise was achieved between an acceptable baseline separation of GalN and GlcN and a baseline separation between GlcN and the huge matrix peak (Fig. 1d).

According to Pohl *et al.* (1997), some functional groups or solutes may change their degree of ionisation at elevated temperatures, which can lead to changes in solute retention and selectivity. It is probable that the matrix peak changed its degree of ionisation, explaining the improved selectivity for this peak at elevated temperatures. Pohl *et al.* (1997) further described the effects of eluent concentrations on the separation. An increase of concentration leads to a decrease of selectivity. For this reason, we decreased the NaOH concentration from 2 mM (Bodé *et al.* 2009) to 1 mM NaOH. However, this change only had a small effect on the separation, compared with the change of temperature, but it improved the separation between matrix peak and GlcN, without impairing the selectivity between GalN and GlcN.

Considering the amino sugar concentrations, there were no significant differences between the new HPAEC method and the HPLC-Fl method. A comparison with results of the literature and the results obtained for the chitin concentration in fungi using the optimised HPAECC-IRMS method showed no significant deviations. We determined an average GlcN concentration of about 20 mg g^{-1} dry matter (Table 2). This is in agreement with chitin concentrations of between 10 to 60 mg g^{-1} dry matter for saprotrophic fungi (Vetter & Siller 1990; Nitschke *et al.* 2011).

5.4.2 The fungal growth experiment

The maize-wood mixtures were preferably decomposed by the fungal species in comparison with the wheat-wood mixture. It was repeatedly shown that decomposition and formation of fungal biomass depend on the substrate quality (Fernandez & Cadish 2003; Kohzu *et al.* 2005; Pant *et al.* 2005). After N addition to the wheat-wood mixture there was initially no difference of the C/N ratio between the two substrates. Therefore, the N concentration and the C/N ratio, respectively, of the substrate may not be responsible for the faster degradation of the maize-wood mixture in the present experiment, contrasting the suggestions by Fernandez and Cadish (2005) and Liang *et al.* (2009).

However, Roger *et al.* (1993) obtained a faster degradation of maize in comparison with wheat substrate by rumen fungi. They explained this observation with the higher hemicellulose concentrations in the maize in comparison with the wheat substrate. Liang *et al.* (2009) tested different plant-wood substrates as a growth medium for *Pleurotus citrinopileatus* and showed a preferred degradation of maize-wood mixture, due to better substrate use efficiency. The formation of GlcN and therefore that of a new fungal biomass is significantly higher in maize-wood than in the wheat-wood mixture (Table 4, Fig.2). This is in line with Leatham (1985), who observed a stimulation of ligninolytic activity by an addition of easily decomposable carbohydrates such as xylose to the wood substrate. However, the newly formed fungal GlcN is maize and not wood derived in the present experiment (Fig. 3).

Several authors (Gleixner et al. 1993; Henn et al. 2004; Kohzu et al. 2005; Hart et al. 2006; Boström et al. 2007; Zeller et al. 2007) observed a $\delta^{13}C$ enrichment in the saprotrophic fungi sporophores of between 1.2 to 6.3‰ relative to the wood decomposed. Zeller et al. (2007) assumed that cellulose and lignin degradation could be involved in $\delta^{13}C$ enrichment of sporcarps of saprotrophic fungi. Kohzu et al. (2005) explained the $\delta^{13}C$ enrichment of saprotrophic fungi relative to their substrate as being due to (1) a selective incorporation of ^{13}C enriched fraction in the substrate by fungi and (2) fungal metabolic processes. They mentioned dark fixation of ambient CO_2 and the kinetic fractionation during assimilation and dissimilation reactions. However, the $\delta^{13}C$ values of the wheat-wood substrate show no $\delta^{13}C$ enrichment. Furthermore, Semenina and Tiunov (2010) showed that there were no significant differences between the shift in $\delta^{13}C$-values of C_4 and C_3 samples during the incubation, using C_4 and C_3 sucrose as substrate.

Consequently, $\delta^{13}C$ fractionation during formation of fungal biomass seems mainly to depend on the substrate. Most likely, the high shifts in the GlcN-specific $\delta^{13}C$ values of maize-wood mixtures were caused by the formation of maize-derived GlcN and a small part might be additionally caused by $\delta^{13}C$ fractionation. The GlcN-specific $\delta^{13}C$ values differed among the fungal species (Table 4), which is in line with Trudell et al. (2004), who observed significant differences between $\delta^{13}C$ values of bulk saprotrophic fungi among genera and species.

5.5 Conclusions

The optimised HPAEC-IRMS method provided reliable results for the concentrations and the $\delta^{13}C$ values of GlcN and GalN in plant samples. The fungal growth experiment indicated that maize-wood mixture was the preferred substrate of saprotrophic fungi in comparison with the wheat-wood mixture. The shifts in $\delta^{13}C$ values were mainly due to the incorporation of maize-derived material into fungal biomass and less due to kinetic isotope fractionation. In the near future, the optimised HPAEC-IRMS method will make it possible to investigate differences in the turnover of, e.g. ecto-mycorrhizal and saprotrophic fungal species in the presence of different organic substrates.

Acknowledgments

We gratefully acknowledge Gabriele Dormann and Reinhard Langel for their technical assistance. Caroline Indorf was funded by the German Research Foundation (DFG).

5.6 References

Appuhn A, Joergensen RG, Scheller E, Wilke B, 2004. The automated determination of glucosamine, galactosamine, muramic acid and mannosamine in soil and root hydrolysates by HPLC. *Journal of Plant Nutrition and Soil Science* **167**: 17–21.

Appuhn A, Joergensen RG, 2006. Microbial colonisation of roots as a function of plant species. *Soil Biology and Biochemistry* **38**: 40-51.

Appuhn A, Scheller E, Joergensen RG, 2006. Relationships between microbial indices in roots and silt loam soils forming a gradient in soil organic matter. *Soil Biology and Biochemistry* **38**: 2557-2564.

Bodé S, Denef K, Boeckx P, 2009. Development and evaluation of a high-performance liquid chromatography isotope ratio mass spectrometry methodology for $\delta^{13}C$ analysis of amino sugars in soil. *Rapid Communications in Mass Spectrometry* **23**: 2519-2526.

Boström B, Comstedt D, Ekblad A, 2008. Can isotope fractionation during respiration explain the ^{13}C-enriched sporocarps of ectomycorrhizal and saprotrophic fungi? *New Phytologist* **177**: 1012-1019.

Cammann K (2001) Allgemeines zur Chromatographie – Wichtige Parameter. In: Cammann K (ed), Instrumentelle Analytische Chemie. Spektrum Akademischer Verlag, Heidelberg; Berlin, pp. 6-7.

Cliquet JB, Deléens E, Bousser A, Martin M, Lescure JC, Prioul L, Mariotti A, Morot-Gaudry JF, 1990. Estimation of carbon and nitrogen allocation during stalk elongation by ^{13}C and ^{15}N tracing in *Zea mays* L. *Plant Physiology* **92**: 79-87.

Decock C, Denef K, Bodé S, Six J, Boeckx P, 2009. Critical assessment of the applicability of gas chromatography-combustion-isotope ratio mass spectrometry to determine amino sugar dynamics in soil. *Rapid Communications in Mass Spectrometry* **23**: 1201-1211.

Dyckmans J, Flessa H, Polle A, Beese F, 2000. The effect of elevated [CO_2] on uptake and allocation of ^{13}C and ^{15}N beech (*Fagus sylvatica* L.) during leafing. *Plant Biology* **2**: 113-120.

Ekblad A, Näsholm T, 1996. Determination of chitin in fungi and mycorrhizal roots by improved

HPLC analysis. *Plant and Soil* **178**: 29-35.

Engelking B, Flessa H, Joergensen RG, 2007a. Microbial use of maize cellulose and sugarcane sucrose monitored by changes in the $^{13}C/^{12}C$ ratio. *Soil Biology and Biochemistry* **39**: 1888-1896.

Engelking B, Flessa H, Joergensen RG, 2007b. Shifts in amino sugar and ergosterol contents after addition of sucrose and cellulose to soil. *Soil Biology and Biochemistry* **39**: 2111-2118.

Engelking B, Flessa H, Joergensen RG, 2008. Formation and use of microbial residues after adding sugarcane sucrose to a heated soil devoid of soil organic matter. *Soil Biology and Biochemistry* **40**: 97-105.

Fernandez I, Cadisch G, 2003. Discrimination against ^{13}C during degradation of simple and complex substrates two white rot fungi. *Rapid Communications in Mass Spectrometry* **17**: 2614–2620.

Gleixner G, Danier HJ, Werner RA, Schmidt HL, 1993. Correlations between ^{13}C content of primary and secondary plant products in different cell compartments and that in decomposing basidiomycetes. *Plant Physiology* **102**: 1287–1290.

Hart SC, Gehring CA, Selmants PC, Deckert RJ, 2006. Carbon and nitrogen elemental and isotopic patterns in macrofungal sporocaps and trees in semiarid forests of the south-western USA. *Functional Ecology* **20**: 42–51.

Henn RM, Chapela IH, 2004. Isotopic fractionation during ammonium assimilation by basidiomycete fungi and its implications for natural nitrogen isotope patterns. *New Phytologist* **162**: 771-781.

Högberg P, Plamboeck AH, Taylor AFS, Fransson PMA, 1999. Natural ^{13}C abundance reveals trophic status of fungi and host-origin of carbon in mycorrhizal fungi in mixed forests. *Proceedings of the National Academy of Sciences of the United States of America* **96**: 8534-8539.

Horvath CG, Preiss BA, Lipsky SR, 1967. Fast liquid chromatography: an investigation of operating parameters and the separation of nucleotides on pellicular ion exchangers. *Analytical Chromatography* **39**: 1422-1428

Indorf C, Dyckmans J, Khan KS, Joergensen RG, 2011. Optimisation of amino sugar quantification by HPLC in soil and plant hydrolysates. *Biology and Fertility of Soils* **47**: 387–396.

Kohzu A, Miyajima T, Tateishi T, Watanabe T, Takashashi M, Wada E, 2005. Dynamics of ^{13}C natural abundance in wood decomposing fungi and their ecophysiological implications. *Soil Biology and Biochemistry* **37**: 1598-1607.

Krummen M, Hilkert AW, Juchelka D, Duhr A, Schlüter H-J, Pesch R, 2004. A new concept for isotope ratio monitoring liquid chromatography / mass spectrometry. *Rapid Communications in Mass Spectrometry* **18**: 2260-2266.

Leathan GF, 1982. Cultivation of shitake, the Japanese forest mushroom, on logs: a potential industry for the United States. *Forest Production Journal* **32** :29-35.

Leatham GF, 1985. Extracellular enzymes produced by the cultivated mushroom *Lentinus edodes* during degradation of a lignocellulosic medium. *Applied and Environmental Microbiology* **50**: 859-867.

Lee RL, Balick MJ, 2006. Flu for you? The common cold, influenza, and traditional medicine. *Explore* **2**: 252-255.

Liang ZC, Wu CY, Shieh ZL, Cheng SL, 2009. Utilization of grass plants for cultivation of *Pleurotus citrinopileatus*. *International Biodeterioration and Biodegradation* **63**: 509-514.

Mannheim T, Braschkat J, Marschner H, 1997. Ammonia emissions from senescing plants and during decomposition of crop residues. *Zeitschrift für Pflanzenernährung und Bodenkunde* **160**: 125-132.

Nitschke J, Altenbach H-J, Malolepszy T, Mölleken H, 2011. A new method for the quantification of chitin and chitosan in edible mushrooms. *Carbohydrate Research* **346**: 1307-1310.

Nursten H, 2005. The chemistry of nonenzymatic browning. In: Nursten H (ed), The Maillard reaction. Chemistry, Biochemistry and Implications. Royal Society of Chemistry, Cambridge, pp 5-30.

Pant D, Reddy UG, Adholeya A, 2006. Cultivation of oyster mushrooms on wheat straw and bagasse substrate amended with distillery effluent. *World Journal of Microbiology and Biotechnology* **22**: 267-275.

Pigman W, Nisizawa K, Tsuiki S, 1959. Chemistry of carbohydrates. *Annual Reviews in Biochemistry* **28**: 15-38.

Pohl CA, Stillian JR, Jackson PE, 1997. Factors controlling ion-exchange selectivity in supressed ion chromatography. *Journal of Chromatography A* **789**: 29-41.

Roger V, Bernalier A, Grenet E, Fonty G, 1993. Degradation of wheat and maize stem by monocentric and a polycentric rumen fungi, alone or in association with rumen cellulolytic bacteria. *Animal Feed Science and Technology* **42**: 69-82.

Semenina EE, Tiunov AV, 2010. Isotopic fractionation by saprotrophic microfungi:

Stamets P, 2000. The role of mushrooms in nature. In: Stamets P (ed), Growing gourmet and medicinal mushrooms. Ten Speed Press, Berkeley California, pp 5-16.

Trudell SA, Rygiewicz PT, Edmonds RL, 2004. Patterns of nitrogen and carbon stable isotope ratios in macrofungi, plants and soils in two old-growth conifer forests. *New Phytologist* **164**:

5. Determination of saprotrophic fungi turnover by glucosamine-specific δ^{13}C LC-IRMS

317-335.

Vetter J, Siller I, 1991. Chitin content of higher fungi. *Zeitschrift für Lebensmittel Untersuchung und Forschung* **193**: 36-38.

Zeller B, Brechet C, Maurice JP, le Tacon F, 2007. ^{13}C and ^{15}N isotopic fractionation in trees, soils and fungi in a natural forest stand and a Norway spruce plantation. *Annals of Forest Science* **64**: 419-429.

6. Zusammenfassung

In der vorliegenden Arbeit wurde eine LC-IRMS Mehode zur aminozucker-spezifischen δ^{13}C-Analyse in Pflanzenmaterialien optimiert und etabliert, um die Bildung und den Umsatz von mikrobiellen Residuen in Boden-und Pflanzenmaterialien mit hoher Genauigkeit erfassen zu können. Weiterhin wurden mit der etablierten Methode zwei Experimente durchgeführt. Der Fokus dieser Arbeit lag jedoch auf der Methodenentwicklung. So wurden verschiedene Aufarbeitungstechniken getestet und optimiert. Außerdem wurde an der HPLC die bereits verfügbare Methode basierend auf Appuhn et al. (2004) optimiert. Zur Etablierung der aminozucker-spezifischen δ^{13}C -Analyse wurden weitere chromatographische Trennmechanismen, mit dem Hintergrund eine möglichst schnelle und effiziente Trennung der Aminozucker mit einem kohlenstofffreien Eluenten zu erzielen, getestet. Weiterhin wurde ein Pilzwachstumsversuch mit saprotrophen Pilzen an Mais-Holz- und Weizen-Holz-Substraten durchgeführt.

Der erste Artikel beschäftigt sich mit der Optimierung einer HPLC-Methode zur Aminozuckerquantifizierung (Appuhn et al., 2004) in Boden- und Pflanzenhydrolysaten. Die Methode von Appuhn et al. (2004) hat zwei Nachteile. (1) Die Methode erbrachte nur zuverlässige Ergebnisse an der Agilent 1100 HPLC und (2) wurde Mannosamin in den meisten Bodenhydrolysaten nur ungenügend getrennt, was zu fehlerhaften hohen Werten in weiteren Publikationen (Appuhn und Jörgensen 2006; Engelking et al. 2007b) zur Folge hatte. Ziel des ersten Artikels war es deshalb die Methode von Appuhn et al. (2004) so zu verbessern, dass an verschiedenen HPLC-Systemen zuverlässige Ergebnisse für alle vier Aminozucker in Boden- und Pflanzenhydrolysaten erhalten werden. Dafür wurde zunächst die mobile Phase optimiert. So wurde der Tetrahydrofurananteil in der mobilen Phase von 0,75% (Appuhn et al., 2004) auf 1,5% erhöht. Durch die Erhöhung des Tetrahydrofurananteils der mobilen Phase konnte deswegen eine kürzere Retentionszeit für alle vier Aminozucker und eine bessere Trennung zwischen Muraminsäure und Mannosamin erzielt werden. Weiterhin wurde die Extinktionswellenlänge des Fluoreszenzdetektors zur Detektion der durch Derivatisierung der Aminozucker erhaltenen Isoindolderivate von 340 nm (Appuhn et al., 2004) auf 330 nm herabgesetzt. Bei dieser Wellenlänge wurde das höchste Signal für die Aminozuckerderivatpeaks detektiert, ohne dass das Signal der Matrixpeaks erhöht wurde. Außerdem hat die OPA-Derivatisierungsreaktionszeit Einfluss auf die Höhe des Fluoreszenzsignals. Bei 120 Sekunden konnte das Maximum für alle vier Aminozucker erzielt werden. Nach Optimierung der genannten Parameter erfolgte die Validierung der Methode. Für Muraminsäure wurde eine Bestimmungsgrenze (LOQ) von 0,5 µmol l^{-1} was 0,13 µg ml^{-1} entspricht. Für die drei anderen Aminozucker wurden 5,0 µmol l^{-1} (entspricht 0,90 µg ml) erhalten. Der lineare Arbeitsbereich lag zwischen 5,0 und 210 µmol l^{-1} (Mannosamin, Galaktosamin und Glucosamin)

6. Zusammenfassung

bzw. zwischen 0,5 und 21 µmol l^{-1} (Muraminsäure). Die relative Standardabweichung betrug durchschnittlich 2% innerhalb eines Tages und 5% innerhalb von sechs Tagen für die Aminozuckerstandards. Weiterhin wurden Wasser und Phosphatpuffer als Probenlösungsmittel getestet, um den Einfluss des pH-Wertes auf die OPA-Reaktion zu testen (Dorresteijn et al., 1996). Es wurde angenommen, dass in Phosphatpuffer gelöste Proben stabiler sind aufgrund des höheren pH-Wertes (pH=5) im Vergleich zu in Wasser gelösten Proben (pH=2-3). Jedoch erzielen die in Wasser aufgenommenen Proben geringere Standardabweichungen, da vermutlich die in der Bodenhydrolysatmatrix enthaltenen Aluminium- und Eisenoxide mit Phosphat Komplexe bilden. Diese Komplexe binden einen Teil der Aminozucker, so dass die gebundenen Aminozucker nicht mehr bestimmt werden können.

Die in Boden- und Pflanzenhydrolysaten mit dieser optimierten HPLC-Methode gemessenen Aminozuckergehalte waren mit den Werten aus der Literatur vergleichbar.

Im zweiten Artikel wurden mehrere HPLC Methoden zur Bestimmung von Aminozuckern in Bodenhydrolysaten getestet und miteinander verglichen. Außerdem sind diverse Aufarbeitungsmethoden zur Purifikation und Konzentrierung der Aminozucker getestet worden. Zur aminozucker-spezifischen $\delta^{13}C$ –Analyse am IRMS ist eine HPLC-Methode mit einer kohlenstofffreien mobilen Phase notwendig, andernfalls kann aufgrund des hohen Hintergrundrauschens kein vernünftiges Signal mehr detektiert werden. Da die im ersten Artikel beschriebene Umkehrphasenmethode einen kohlenstoffhaltigen Eluenten enthält, musste eine ebenso zuverlässige Methode, die jedoch keine organischen Lösungsmittel benötigt, getestet und mit der schon etablierten Methode verglichen werden. Es gibt eine Reihe von HPLC-Methoden, die ohne organische Lösungsmittel auskommen, wie z. B. (1) HPAEC, (2) Hochleistungskationenaustauschchromatographie (HPCEC) und (3) die Hochleistungsanionenausschlusschromatographie (HPEXC). Die von Bodé et al. (2009) beschriebene HPAEC- Methode für die Isotopenaminozuckeranalyse hat einige Nachteile: (1) Das NaOH in der mobilen Phase reagiert schnell mit CO_2 aus der Luft zu Carbonat und führt so zu einem hohen Hintergrundrauschen und verschlechtert die Trennung der Aminozucker auf der Analysensäule, (2) die alkalische mobile Phase setzt im Idealfall ein metallfreies System voraus, (3) die von Bodé et al. (2009) beschriebene Methode besteht aus zwei separaten Läufen (Muraminsäure und basische Aminozucker), und (4) die Nachweisgrenze von Muraminsäure ist zu hoch für die geringe Muraminsäurekonzentration in den Probenhydrolysaten. Aufgrund dieser Faktoren waren die Ziele der Studie (1) eine zuverlässige Purifikations- und Konzentrierungsmethode für Aminozucker in HCl-Hydrolysaten zu finden und (2) die Methode von Bodé et al. (2009) zu optimieren durch das Testen verschiedener Trennungsmechanismen. Es wurden folgende fünf Aufarbeitungsmethoden zur Purifikation und Konzentrierung der Probenhydrolysate getestet: (1)

6. Zusammenfassung

Extraktion basierend auf Appuhn (2004), (2) Extraktion basierend auf Zhang und Amelung (1996), (3) Extraktion mittels eines starken Kationenaustauscherharzes auf Polymerbasis, (4) Extraktion mittels eines starken Kationenaustauscherharzes auf Silicabasis und (5) Extraktion mittels eines starken Anionenaustauscherharzes zur Konzentrierung und Separation von Muraminsäure. Die besten Ergebnisse wurden mit den Kationenaustauscherharzen erzielt, wobei die dritte Methode noch ein wenig besser geeignet war als die vierte. Nach Anwendung der dritten Methode konnten die Matrixeinflüsse leicht reduziert und die Aminozucker konzentriert werden. Zur weiteren Trennung der Aminozucker mittels HPLC konnte für die HPCEC und die HPEXC keine zufriedenstellende Trennung der Aminozucker erreicht werden. Die von Bodé et al. (2009) beschriebene HPAEC-Methode erwies sich als geeignet und es konnten mit der Umkehrphasenmethode vergleichbare Ergebnisse für Bodenhydrolysate erzielt werden. Die HPAEC-Methode wurde zunächst mittels OPA-Nachsäulenderivatisierung und Fluoreszenzdetektion durchgeführt. Bei der Muraminsäuremethode wurde als Eluent Natriumnitrat durch Natriumacetat ersetzt, da Nitrat die Fluoreszenz gelöscht hat. Aufgrund der mittels Fluoreszenzdetektion erhaltenen breiten Matrixpeaks, konnte bei dieser Methode kein Gradient angewendet werden, so dass Muraminsäure in einem separaten Lauf bestimmt werden musste. Weiterhin wurde ein elektrochemischer Detektor zur Detektion getestet. Hier waren die Peaks schmaler, so dass es möglich war alle vier Aminozucker mit Hilfe eines Laufmittelgradienten in einem Lauf zu bestimmen. Schlussfolgernd kann zusammengefasst werden, dass für Detektoren mit geringer Empfindlickeit (z.B. IRMS) eine Konzentrierung und Purifikation insbesondere von Muraminsäure über ein Kationenaustauscherharz sinnvoll ist. Eine Basislinientrennung für alle Aminozucker war nur mit der HPAEC möglich. Da mit dieser Methode gute Validierungdaten erzielt wurden und die Aminozuckergehalte mit der Umkehrphasenmethode vergleichbar waren, stellt die HPAEC die Methode der Wahl zur aminozucker-spezifischen $\delta^{13}C$–Analyse am IRMS dar.

Der dritte Artikel befasst sich mit der Optimierung der aminozucker-spezifischen $\delta^{13}C$ – Analyse mittels HPAEC-IRMS in Pflanzenhydrolysaten sowie mit der Bestimmung des Umsatzes von saprotrophen Pilzen in verschieden Substraten. Die von Bodé et al. (2009) beschriebene HPAEC-IRMS- Methode ist für die aminozucker-spezifische $\delta^{13}C$–Analyse in Bodenhydrolysaten jedoch nicht in Pflanzenhydrolysaten geeignet. In Pflanzenhydrolysaten wird der Glucosaminpeak von Peaks aus der Matrix interferiert. Folglich war das erste Ziel dieses Artikels, die Methode so zu optimieren, dass eine aminozucker-spezifische $\delta^{13}C$ –Analyse in Pflanzenhydrolysaten möglich ist. Weiterhin sollten mit der optimierten HPAEC-IRMS-Methode die Bildung und der Umsatz von saprotrophen Pilzen bestimmt werden. Durch Änderung der Säulentemperatur von 15 °C auf 38 °C und durch Herabsetzung der NaOH- Konzentration von 2 mM auf 1 mM konnte der störende

6. Zusammenfassung

Matrixpeak aus dem Pflanzenhydrolysat von dem Glucosaminpeak basisliniengetrennt werden. Die Validierungsparameter waren gut und die bestimmten Aminozuckergehalte waren mit der Umkehrphasen-HPLC-Methode vergleichbar. Zur Bestimmung der Bildung und des Umsatzes von saprotrophen Pilzen auf verschiedenen Substraten wurden *Lentinula edodes* P., *Pleurotus ostreatus* K. und *Pleurotus citrinopileatus* S. auf Mais-Holz- und auf Weizen-Holz-Substrat für vier Wochen bei 24 °C kultiviert. Dieser Pilzwachstumsversuch zeigte, dass 80% des neu gebildeten pilzlichen Glucusamins maisbürtig und nicht holzbürtig waren. Weiterhin wurde der bevorzugte Abbau von Maissubstrat im Vergleich zu Weizensubstrat an diesem Versuch verdeutlicht. Außerdem lassen die Ergebnisse darauf schließen, dass die beobachtete zunehmende $\delta^{13}C$ Anreicherung in dem neu gebildeten pilzlichen Glucosamin während der vier Wochen auf die Inkorporation des angereicherten ^{13}C aus dem Substrat und eher weniger auf kinetische Isotopeneffekte zurückzuführen ist.

7. Summary

In the present thesis a method for amino sugar- specific $\delta^{13}C$–analysis via LC-IRMS in plant material was optimised and established for determining the formation and turnover of microbial residues in soil- and plant hydrolysates with high accuracy. Further an experiment was implemented by the established LC-IRMS method. However, the main focus of the thesis was on method development. Thus, different pretreatment techniques were tested and optimised. Moreover, the already available HPLC method based on Appuhn et al. (2004) was optimised. For establishing the amino sugar- specific $\delta^{13}C$–analysis additional chromatographic separation mechanisms were tested with the background to yield a fast and efficient separation of amino sugars with a carbon-free eluent.

The implemented experiment was a fungi growth experiment with saprotrophic fungi on maize-wood- and wheat-wood-substrates.

The first article deals with the optimisation of amino sugar quantification by HPLC in soil and plant hydrolysates. The already available HPLC method based on Appuhn et al. (2004) was hampered by two drawbacks: (1) The method was reliably working on the Agilent 1100 HPLC equipment and (2) mannosamine was insufficiently separated in most soil hydrolysates, which resulted in erroneous high values usually omitted in further publications (Appuhn and Jörgensen, 2006; Engelking et al., 2007b). The objective of the first article was improving the method of Appuhn et al. (2004) to give reliable results for all four amino sugars in soil and plant hydrolysates using different HPLC equipments. At first the mobile phase was optimized. Therfore the tetrahydrofurane (THF) concentration was increased from 0.75% to 1.5%. This yielded in shorter retention times for all four amino sugars and a slightly better resolution between muramic acid and mannosamine. Moreover, the excitation wavelength of the fluorescence detector was reduced from 340 nm (Appuhn et al. 2004) to 330 nm. At 330 nm the highest fluorescence response for the amino sugar peaks was detected without increasing the fluorescence response of the matrixpeaks. Besides the OPA reaction time has an effect on the fluorescence response. A maximum height of fluorescence response for all four amino sugars was yielded at 120 sec. After optimisation of the mentioned parameters the validation of the method ensued. A limit of quantification (LOQ) of 0.5 µmol l^{-1} which is equal to 0.13 µg ml $^{-1}$ for muramic acid and a LOQ of 5.0 µmol l^{-1} (equal to 0.90 mg ml $^{-1}$) for the other three amino sugars were obtained. The calibration curves were linear in the range from 5.0 to 210 µmol l^{-1} (mannosamine, galactosamine, glucosamine) and from 0.5 to 21 µmol l^{-1} (muramic acid), respectively. The coefficient of variation was roughly 2% for intraday and 5% for interday precision, respectively. Furthermore water and phosphate buffer as sample solvent were tested, because the pH value of the sample solvent has an effect of the OPA reaction (Dorresteijn et al.,

1996). Since sample pH is between 2 and 3 in the sample extract using water, it was examined if better accuracy for this method is possible with higher pH during OPA reaction. Therefore the sample extract pH was increased to values about 5 with phosphate buffer. However, buffered samples showed higher standard deviations, particularly for the basalt-derived clayey soil forest, which contained high contents of aluminium and iron oxides. The hydrolysis products of these oxides presumably form as central ion complexes in the presence of ligands like phosphate. If these complexes bind amino sugars, they may not be completely available for OPA derivatisation anymore.

The amino sugar contents in soil- and planthydrolysates measured by this optimised HPLC method compare well with those reported by the literature from arable and forest soils.

In the second article HPLC methods for the determination of amino sugars in soil hydrolysates were tested and compared. Moreover various pretreatment methods for purification and concentration of amino sugars were tested. For amino sugar-specific $\delta^{13}C$ analysis a HPLC method with a carbon-free eluent is necessary, otherwise a high background noise would be occurred. The reversed phase HPLC method (mentioned in the first article) is reliable, but is not suited for adaption with IRMS, due to the organic mobile phase. However, a variety of other HPLC methods are free of organic solvents, such as (1) HPAEC, (2) high performance cation exchange chromatography (HPCEC) and (3) high performance anion exclusion chromatography (HPEXC). The already available HPAEC method for amino sugar isotope analysis (Bodé et al., 2009) has several drawbacks: (1) The NaOH of the mobile phase is sensitive to CO_2 contamination, (2) the alkaline mobile phase ideally requires a metal free liquid handling system (3) the method of Bodé et al. consisted of two separate runs, one for the basic amino sugars and one for muramic acid, and (4) the limit of quantification for muramic acid is too high for the low muramic acid concentrations in soil. For these reasons the objectives of the second article were (1) to find a reliable purification and concentration procedure for amino sugars in HCl hydrolysates, (2) to optimise the method of Bodé et al. (2009) by testing different chromatographic separation mechanisms. To have higher amino sugar concentration and less impurity in the soil hydrolysates, the following amino sugar purification methods were tested: (1) extraction based on Appuhn et al. (2004), (2) extraction based on Zhang and Amelung (1996), (3) extraction via polymer based strong cation exchange resin, (4) extraction via silica based strong cation exchange resin and (5) extraction via strong anion exchange resin to concentrate and separate muramic acid from the basic amino sugars. The strong cation exchange purification methods gave the highest amounts for all four amino sugars, without an increase of impurities, whereas the polymer based cation exchange purification method (3) gave slightly higher amount than the silica based purification method (4). For the further separation of amino sugars no satisfactory separation with the HPCEC and the HPEXC mechanisms was obtained. The HPAEC method described by Bodé et al. (2009) showed a baseline separation and the determined amino sugar amounts in soil hydrolysates were comparable with the reversed phase HPLC method described in the first article. First the HPAEC method was implemented via post column OPA derivatisation and fluorescence detection. For the determination of muramic acid sodium acetate was substituted for sodium nitrate as component of the mobile phase because nitrate acted as a quencher and affected the fluorescence detection. The use of a gradient for obtaining muramic acid together with the basic amino sugars in one run was not possible, due to the fact that muramic acid was overlayed by peaks of the matrix. Additionally a pulsed amperometric electrochemical detector (PAD) was used for detection. The advantage of the PAD is the lack of derivatisation. The obtained peaks were smaller and therefore the peaks were not interfered by peaks of the matrix. Thus, the use of a gradient was possible and therefore the four amino sugars were determined in one run. In summary, a concentration and purification particularly of muramic acid is recommended for detection methods with low sensitivity. A baseline separation for the amino sugars was only obtained by the HPAEC mechanism. Good validation data were yielded with the HPAEC method and the amino sugar amount in the measured soil hydrolysates were comparable with those determined via reversed phase HPLC. Therefore the HPAEC mechanism is at the moment the best choice for amino sugar- specific $\delta^{13}C$ analysis.

The third article deals with the optimisation of amino sugar-specific $\delta^{13}C$ analysis via HPAEC-IRMS in plant hydrolysates as well as with the determination of saprotrophic fungi turnover in different substrates. The HPAEC-IRMS method described by Bodé et al. (2009) performed well for soil samples but for the plant material the matrix of the plant samples interfered with glucosamine. Consequently the first objective was to adapt the HPAEC-IRMS method described by Bodé et al. (2009) to an amino sugar analysis in plant and fungi samples. A further objective was the determination of the formation and turnover of saprotrophic fungi, particularly the substrate of the newly formed fungal biomass by the optimized HPAEC-IRMS. By increasing the column temperature from 15 °C to 38 °C and by decreasing the NaOH concentration of the eluent from 2 mM to 1 mM the interfering peak of the matrix was separated from the glucosamine peak. The validation data were good and measured amino sugar amounts were comparable with those determined by reversed phase HPLC.

For determining the formation and the turnover of saprotrophic fungi on different substrates *Lentinula edodes* P., *Pleurotus ostreatus* K. und *Pleurotus citrinopileatus* S. were cultivated on maize-wood- and wheat-wood-substrates for four weeks at 24 °C. The fungi growth experiment showed that 80% of the new formed fungal glucosamine was maize-derived and not wood-derived. Furthermore this experiment indicated the preferred decomposition of maize substrate in comparison to wheat substrate. Also, the results showed that the increasing $\delta^{13}C$ enrichment in the

7. Summary

new formed fungal glucosamine during the incubation stems from the incorporation of the enriched ^{13}C of the substrate und less from kinetic isotope effects.

8. Schlussfolgerung und Ausblick

Die Ergebnisse dieser Arbeit haben gezeigt, dass die aminozucker-spezifische $\delta^{13}C$ Analyse in Pflanzen- und Bodenhydrolysaten mittels LC-IRMS möglich ist und eine höhere Genauigkeit im Vergleich zum GC-C-IRMS aufweist. So ist es mit der GC-C-IRMS nur möglich $\delta^{13}C$ Veränderungen die größer als 2 ‰ sind genau zu bestimmen (Glaser und Gross, 2005). Mit der in dieser Arbeit beschriebenen LC-IRMS Methode können auch kleine Veränderungen des $\delta^{13}C$ im Boden präzise bestimmt werden. Die Standardabweichungen der Matrixkalibrierung (mit Boden gespikte Proben) lagen unterhalb von 0,35 ‰ für die $\delta^{13}C$-Werte. Durch diese erlangte Genauigkeit kann bei Anwendung dieser Methode zwischen frisch gebildeten mikrobiellen Residuen und älterem sich im Boden akkumuliertem Material mikrobiellen Ursprungs unterschieden werden. Die LC-IRMS Methode erlaubt es auch, Unterschiede zwischen den Umsätzen von bakteriellen und pilzlichen Residuen zu bestimmen. Weiterhin wäre die Methode hilfreich um die Bedeutung von Galaktosamin zu klären. Da der Galaktosamingehalt während des von Engelking et al. (2007b) durchgeführten Inkubationsversuchs keine Veränderungen zeigte, bietet LC-IRMS die Möglichkeit zu klären, ob die Mikroorganismen kein Galactosamin produzieren, oder ob der Abbau und die neue Bildung von Galaktosamin gleich groß sind. Darüber hinaus ist die Methode dafür geeignet viele Hypothesen zu klären. Eine davon ist beispielsweise:

(1) Die Bildung von mikrobiellen Residuen aus C4-Saccharose ist abhängig von der Qualität der mikrobiellen Residuen und der mikrobiellen Gemeinschaftsstruktur im Boden. Es werden in Böden mit einem hohen Gehalt an pilzlichem Glucosamin mehr mikrobielle Residuen gebildet.

Außerdem besteht die Aussicht mittels LC-IRMS den Einfluss von N auf die Bildung der mikrobiellen Residuen zu bestimmen. So kann in Zukunft Fragestellungen nachgegangen werden, ob nach Zugabe von ^{13}C markiertem Streu mehr pilzliche als bakterielle Residuen bei einem sinkenden N Gehalt und einem zunehmenden Ligningehalt gebildet werden. Weiterhin ist es möglich, den Einfluss der Temperatur auf die Bildung mikrobieller Residuen zu bestimmen. Wird beispielsweise bei niedrigen Temperaturen mehr Substrat-C in die mikrobielle Biomasse und weniger in die mikrobiellen Residuen eingebaut?

Die im ersten Artikel optimierte Umkehrphasen-HPLC-Methode zur Aminozuckerbestimmung ist relativ schnell und einfach durchzuführen und liefert zuverlässige Ergebnisse. Somit sollte es kein Problem darstellen, diese Methode an verschiedenen HPLC-Anlagen durchzuführen, vorausgesetzt es ist ein für die Online-Derivatisierung vorgesehener Probengeber vorhanden. Weiterhin sollte die im ersten Artikel beschriebene Aufarbeitung den jeweiligen Proben angepasst werden. So sollte bei Böden mit sehr hohem Carbonatgehalt das nach der Hydrolyse erhaltene

8. Schlussfolgerung und Ausblick

Hydrolysat vollständig zur Trockne eingeengt werden und nach Aufnahme in Wasser neutralisiert werden, damit die Carbonate gefällt werden. Diese könnten ansonsten bei der Derivatisierung stören.

Die im zweiten Artikel vorgestellten Ergebnisse, zeigen dass die HPAEC- Methode zuverlässig für die aminozucker spezifische ^{13}C –Analyse in Bodenhydrolysaten ist. Falls keine hohen Mengen an Mannosamin (Nachweisgrenze: 50 µmol l^{-1} Bodenextrakt) in Böden enthalten sind, ist die im dritten Artikel vorgestellte optimierte HPAEC-IRMS- Methode vorzuziehen, da es mit dieser Methode möglich ist neben Bodenhydrolysaten auch Pflanzenhydrolysate bei einer kürzeren Retentionszeit zu messen. In den meisten Bodenextrakten ist der Mannosamingehalt mit 0-20 µmol l^{-1} Bodenextrakt weit unter der Nachweisgrenze der HPAEC-IRMS Methode. Die im zweiten Artikel beschriebene Aufarbeitungsmethode zur Purifikation und Konzentrierung der Aminozucker ist insbesondere für die Konzentrierung der basischen Aminozucker erfolgreich. Weiterer Forschungsbedarf besteht aber bei der Konzentrierung von Muraminsäure, so dass nicht, wie im dritten Kapitel beschrieben, die Probenextrakte nach vollständiger Aufarbeitung noch zusätzlich durch Einengung konzentriert werden müssen. Ein nächster Versuch sollte sein die bereits sauren pH (2-3) Bodenextrakte vor Verwendung des im zweiten Artikel beschriebenen Kationenaustauscherharzes auf einen pH-Wert von <1 einzustellen. Theoretisch liegt bei diesem pH-Wert Muraminsäure als diprotoniertes Kation (siehe Einleitung) vor und dürfte erst nach Zugabe der Salzsäure von dem Austauscherharz eluieren. Bei einem pH von 2-3 kann die protonierte Aminogruppe zwar anziehend, die deprotonierte Carboxylgruppe jedoch abstoßend auf die Sulfonsäuregruppe des Harzes reagieren, so dass ein Teil der Muraminsäurefraktion schon während des Waschens mit Wasser von dem Harz eluiert.

Die im dritten Artikel optimierte LC-IRMS Methode lieferte sehr zuverlässige Ergebnisse für die Gehalte so wie auch für die δ^{13}C-Werte von Galaktosamin und Glucosamin in Pflanzen-und Bodenhydrolysaten. Die Methode kann somit zur Bestimmung von Galaktosamin und Glucosamin für Pflanzen- und Bodenproben angewandt werden. Der Pilzwachstumsversuch hat verdeutlicht, dass die ^{13}C Anreicherung in dem neu gebildeten pilzlichen Glucosamin auf die Inkorporation von maisbürtigem Material und weniger auf kinetische Isotopenfraktionierung zurückzuführen ist. Der Versuch hat außerdem gezeigt, dass die saprotrophen Pilze besonders schnell auf Mais-Holz-Substrat wachsen und 80% des neuen pilzlichen Glucosamin maisbürtig sind. Daraus folgt, dass für eine schnelle Kultivierung von saprotrophen Pilzen Mais als Substrat verwendet werden sollte. Weiterhin könnten mit dieser Methode Unterschiede zwischen den Umsätzen von saprotrophen Pilzen und Ektomykorrhiza auf verschiedenen Substraten bestimmt und beurteilt werden.

9. Literatur

Ågren, GI, Bosatta E, Balesdent, J, 1996. Isotope discrimination during decomposition of organic matter. A theoretical analysis. Soil science Society of America Journal 60: 1121-1126.

Amelung W, Zhang X, Flach KW Zech W, 1999. Amino sugars in native grassland soils along a climosequence in North America. Soil Science Society of America Journal 63: 86-92.

Amelung W, 2001. Methods using amino sugars as markers for microbial residues in soil. In: Lal, J.M., Follett, R.F., Stewart, B.A. (Eds.) Assessment methods for soil carbon. Lewis Publishers, Boca Raton, pp.233-272.

Amelung W, Lobe I, Du Preez CC, 2002. Fate of microbial residues in sandy soils of the South African Highveld as influenced by prolonged arable cropping. European Journal of Soil Science 53, 29-35.

Amelung W, 2003. Nitrogen biomarkers and their fate in soil. Journal of Plant Nutrition & Soil Science 166, 677-686.

Amelung W, Brodowski S, Sandhage-Hofmann A, Bol R, 2008. Combining biomarker with stable isotope analyses for assessing the transformation and turnover of soil organic matter. Adv Agron 100:155-250

Appuhn A, Joergensen RG, Raubuch M, Scheller E, Wilke B, 2004. The automated determination of glucosamine, galactosamine, muramic acid and mannosamine in soil and root hydrolysates by HPLC. Journal of Plant Nutrition & Soil Science 167, 17-21.

Appuhn A, Joergensen RG, 2006. Microbial colonisation of roots as a function of plant species. Soil Biology & Biochemistry 38, 1040-1051.

Balota EL, Colozzi-Filho A, Andrade DS, Dick RP, 2003. Microbial biomass in soils under different tillage and crop rotation systems. Biol. Fertil. Soils 38: 15.

Bodé S, Denef K, Boeckx P, 2009. Development and evaluation of a high-performance liquid chromatography isotope ratio mass spectrometry methodology for $\delta^{13}C$ analysis of amino sugars in soil. Rapid Comm. Mass Spec.. 23, 2519-2526.

Bondietti E, Martin JP, Haider K, 1972. Stabilization of amino sugar units in humic-type polymers. Soil Science Society of America Proceedings 36, 597-602.

Boutton TW, 1991. Stable carbon isotope ratios of natural materials: I. Sample Preparation and Mass Spectrometric Analysis. In: Carbon Isotope Techniques (Coleman DC, Fry B, Eds.), Academic Press Inc. California, 155-171.

Cammann K, 2001. Allgemeines zur Chromatographie – Wichtige Parameter. In: Cammann K (ed), Instrumentelle Analytische Chemie. Spektrum Akademischer Verlag, Heidelberg; Berlin, pp. 6-7.

9. Literatur

Carle R, 1991. Isotopen-Massenspektrometrie – Grundlagen und Anwendungsmöglichkeiten. Pharmazie in unserer Zeit. 20 (2) : 75-82.

Coelho RRR, Sacramento DR, Linhares LF, 1997. Amino sugars in fungal melanins and soil humic acids. Eur J Soil Sci 48:425-429

Collins HP, Blevins RL, Bundy LG, Christenson DR, Dick WA, Huggins DR, Paul EA, 1999. Soil carbon dynamics in corn-based agroecosystems: results from carbon-13 natural abundance. Soil Society of America Journal. 63: 584-591.

Dorresteijn RC, Berwald LG, Zomer G, de Gooijer CD, Wieten G, Beuvery EC, 1996. Determination of amino acids using o-phthalaldehyde-2-mercaptoethanol derivatization. Effect of reaction conditions. J Chromatogr A 724:159–167

Engelking B, Flessa H, Joergensen RG, 2007a. Microbial use of maize cellulose and sugarcane sucrose monitored by changes in the $^{13}C/^{12}C$ ratio. Soil Biology and Biochemistry 39: 1888-1896.

Engelking B, Flessa H, Joergensen RG, 2007b. Shifts in amino sugar and ergosterol contents after addition of sucrose and cellulose to soil. Soil Biology and Biochemistry 39: 2111-2118.

Ferrero MÁ, Aparicio LR, 2010. Biosynthesis and production of polysialic acids in bacteria. Appl Microbiol Biotechnol 86:1621-1635

Glaser B, Gross S, 2005. Compound-specific $\delta^{13}C$ analysis of individual amino sugars - a tool to quantify timing and amount of soil microbial residue stabilization. Rapid Comm. Mass Spec. 19, 1409-1416.

Goh KM, 1991. Carbon dating. In: Coleman DC and Fry B (eds), Carbon Isotope Techniques. Academic Press, Santiego Isotope Techniques in Plant, Soil and Aquatic Biology, pp. 125-145.

Gregorich EG, Ellert BH, Drury CF, Liang BC, 1996. Fertilization effects on soil organic matter turnover and corn residue C storage. Soil Science Society of America Journal 60: 472-476.

Guggenberger G, Frey SD, Six J, Paustian K, Elliott ET, 1999. Bacterial and fungal cell wall residues in conventional and no-tillage agroecosystems. Soil Sci Soc Am J 63:1188–1198

IPCC, 2007. Coupling Between Changes in the Climate System and Biogeochemistry. In: Solomon S, Quin M, Manning Z, Chen M, Marquis KB, Averyt K, Tignor M, Miller HL (eds), Climate Change 2007: a Physical science Basis, Camebridge University Press: Camebridge, 499.

Joergensen RG, Meyer B, 1990. Chemical change in organic matter decomposing in and on a forest Rendzina under beech (Fagus sylvatica L.). J Soil Sci 41:17-27.

Kenne LK, Lindburg B, 1983. Bacterial polysaccharides. In: Aspinall GO (ed) The polysaccharides. Academic Press, New York, pp 287-353.

9. Literatur

Liang C, Zhang X, Rubert KF, Balser TC, 2006. Effect of plant materials on microbial transformation of amino sugars in three soil microcosms. Biol Fertil Soils 43:631-639

Liang C, Zhang X, Balser TC, 2007a. Net microbial amino sugar accumulation process in soil as influenced by different plant material inputs. Biol Fertil Soils 44:1-7.

Liang C, Fujinuma R, Wei LP, Balser TC, 2007b. Tree species-specific effects on soil microbial residues in an upper Michigan old-growth forest system. Forestry 80:65-72.

Ludwig B, John B, Ellerbrock R, Kaiser M, Flessa H, 2003. Stabilization of carbon from maize in a sandy soil in a long-term experiment. European Journal of Soil Science 54: 117-126.

Millar WN, Casida LE, 1970. Evidence for muramic acid in soil. Can J Microbiol 16:299-304.

Monsandl A, Henner U, Fuchs S, 2000. Natürliche Duft- und Aromastoffe- Echtheitsbewertung mittels enantioselektiver Kapillar-GC und/oder Isotopenverhältnismassenspektrometrie. In: Günzler H (ed), Analytiker Taschenbuch Bd.21, Springer Verlag, Berlin, pp. 53-62.

Nultsch W, 2001. Energieumwandlungen und Syntheseleistungen. In: Nultsch W, (Eds.), Allgemeine Botanik, Volume 11.Georg Thieme, Stuttgart, pp. 329-332.

Parsons, JW, 1981. Chemistry and distribution of amino sugars in soils and soil organisms. In: Paul EA, Ladd JN (Eds.), Soil Biochemistry, Volume 5. Marcel Dekker, New York, pp. 197-227.

Scholle G, Joergensen RG, Schaefer M, Wolters V, 1993. Hexosamines in the organic layer of two beech forest soils: effects of mesofauna exclusion. Biology & Fertility of Soils 15, 301-307.

Sharon N, 1965. Distribution of amino sugars in microoganisms, plants and invertebrates. In: Balasz EA, Jeanlanx RW (eds) The amino sugars, part 2A. Distribution and biological role. Academic Press, New York, pp 1-45.

Stevenson F.J, 1982. Organic forms of soil nitrogen. In: Stevenson FJ (ed) Nitrogen in Agricultural Soils. American Society of Agronomy, Madison, FL, pp 101-104.

Ternes W, Täufel A, Tunger L, Zobel M, 2007. In: Lexikon der Lebensmittelchemie. Wissenschaftliche Verlagsgesellschaft, Stuttgart, pp 74-79.

Vollhardt KPC, Schore NE, 2000. Aminosäuren, Peptide und Proteine. In: Vollhardt KPC, Schore NE (eds) Organische Chemie. Willey-VCH, Weinheim, pp. 1288-1289

Wasylnka JA, Simmer MI, Moore MM, 2001. Differences in sialic acid density in pathogenic and non-pathogenic Aspergillus species. Microbiol 147:869-877.

Wu J, O'Donnell G, Syers JK, 1993. Microbial growth and sulphur immobilization following the incorporation of plant residues into soil. Soil Biol. Biochem. 25: 1567-1573.

Yoneyama T, Koike, Y, Arakawa Y, Yokoyama HK, Sasaki Y, Kawamura T, Araki Y, Ito E, Takao S, 1982. Distribution of mannosamine and mannosaminuronic acid among cell walls of Bacillus species. J Bacteriol 149:15-21.

Zelles L, Alef K, 1995. Biomarkers. In: Alef, K., Nannipieri, P. (Eds.), Methods in Applied Soil Microbiology and Biochemistry. Academic Press, London. pp. 422-439.

Zhang X, Amelung W, 1996. Gas chromatographic determination of muramic acid, glucosamine, mannosamine, and galactosamine in soils. Soil Biol. Biochem. 28, 1201–1206.

i want morebooks!

Buy your books fast and straightforward online - at one of world's fastest growing online book stores! Environmentally sound due to Print-on-Demand technologies.

Buy your books online at
www.get-morebooks.com

Kaufen Sie Ihre Bücher schnell und unkompliziert online – auf einer der am schnellsten wachsenden Buchhandelsplattformen weltweit! Dank Print-On-Demand umwelt- und ressourcenschonend produziert.

Bücher schneller online kaufen
www.morebooks.de

 VDM Verlagsservicegesellschaft mbH
Heinrich-Böcking-Str. 6-8　　Telefon: +49 681 3720 174　　info@vdm-vsg.de
D - 66121 Saarbrücken　　　Telefax: +49 681 3720 1749　　www.vdm-vsg.de

Printed by Books on Demand GmbH, Norderstedt / Germany